CAPITAL PRESERVATION

Preparing for Urban Operations in the Twenty-First Century

Proceedings of the RAND Arroyo-TRADOC-MCWL-OSD Urban Operations Conference, March 22–23, 2000

Edited by
Russell W. Glenn

RAND
ARROYO CENTER

The research described in this report was sponsored by the United States Army under Contract No. DASW01-96-C-0004.

ISBN: 0-8330-3008-6

Cover artwork by Priscilla B. Glenn
Cover design by Tanya Maiboroda

Published 2001 by RAND
1700 Main Street, P.O. Box 2138, Santa Monica, CA 90407-2138
1200 South Hayes Street, Arlington, VA 22202-5050
201 North Craig Street, Suite 102, Pittsburgh, PA 15213
RAND URL: http://www.rand.org/
To order RAND documents or to obtain additional information, contact Distribution Services: Telephone: (310) 451-7002;
Fax: (310) 451-6915; Email: order@rand.org

PREFACE

On March 22–23, 2000, the RAND Arroyo Center, U.S. Army Training and Doctrine Command (TRADOC), Marine Corps Warfighting Lab (MCWL), and Office of the Secretary of Defense co-hosted a conference on military urban operations at RAND's headquarters in Santa Monica, California. The conference sought to provide a forum for education and debate on current and future challenges inherent in the commitment of armed forces to the world's villages, towns, and cities. The conference agenda, with a list of speakers and panel members, appears in Appendix A.

This conference proceedings is part of a larger RAND Arroyo Center effort to identify U.S. force requirements in preparing for urban contingencies and to develop innovative approaches to meeting the challenges inherent in such undertakings. It will be of interest to government and commercial-sector personnel whose responsibilities include policy design, funding, planning, preparation, or the development of technologies for urban operations.

The research is sponsored by the Deputy Assistant Secretary of the Army for Research and Technology and is being conducted within RAND Arroyo Center's Force Development and Technology Program. The Arroyo Center is a federally funded research and development center sponsored by the United States Army.

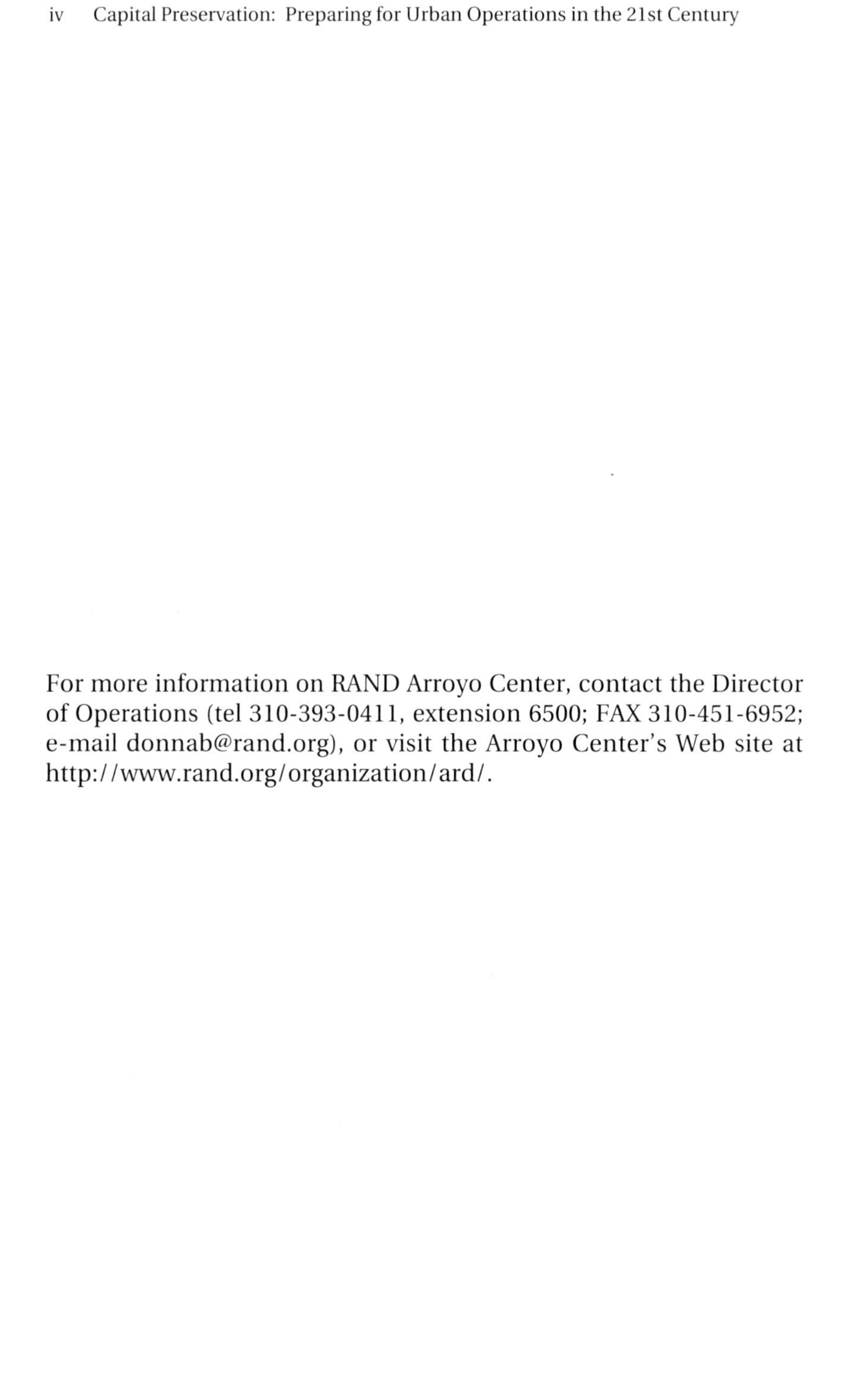

For more information on RAND Arroyo Center, contact the Director of Operations (tel 310-393-0411, extension 6500; FAX 310-451-6952; e-mail donnab@rand.org), or visit the Arroyo Center's Web site at http://www.rand.org/organization/ard/.

CONTENTS

SUMMARY

The 2000 Urban Operations Conference was held in Santa Monica, California, and hosted by the RAND Arroyo Center, U.S. Army Training and Doctrine Command, Marine Corps Warfighting Lab, and the Office of the Secretary of Defense. The objectives of the event were to:

- Explain the significance of urban areas to current and future military operations.
- Consider and discuss methods and means of seizing, stabilizing, or controlling urban areas in the 21st century.
- Identify technology requirements across the spectrum of urban operations.
- Identify C4ISR requirements inherent in military urban operations and ways of meeting those requirements.

This conference proceedings includes an introduction and annotated copies of the briefings given by each speaker.

After Dr. David Chu's opening of the conference, the first two lectures considered the issue of urban combat in Chechnya. **General Anatoly Sergeevich Kulikov**, commander of Russian forces in Chechnya between February and July 1995, spoke on the challenges confronted by his nation's military during urban operations in both the 1994–1996 and 1999–2000 campaigns and the lessons learned from each. He was followed by **Arthur L. Speyer, III**, of the Marine Corps Intelligence Agency, who provided a Western analysis of both Russian and Chechen urban operations during those campaigns.

Moving from the subject of war to those of stability missions and nation building, **LtCol Philip J. Gibbons** addressed his nation of New Zealand's participation in East Timor operations and the role of built-up areas in those actions. **LtCol John M. Allison (USMC, ret.)** was responsible for force protection during the United States Marine Corps early actions in Somalia. His lecture provided insights into that area of perpetual concern, a subject area seldom addressed in any detail in the urban operations literature. Two other officers with considerable experience in dealing with urban areas during U.S. stability missions in Bosnia-Herzegovina opened the first day's afternoon sessions. **Colonel Greg Fontenot** was the United States' first brigade commander to assume responsibility for the American sector that included the strategic city of Brcko and its crossings over the Sava River. His comments on the need to balance military force with other, nonmilitary elements of power demonstrated the complex nature of urban operations and the criticality of integrating urban-related initiatives with those involving the rural surroundings of which they are a part. **Colonel James K. Greer** commanded in the same region as a battalion commander, having done so during the Serb riots of August 1997. His observations into the tactical and operational repercussions of those events reflected the broad scope of demands for which a unit has to plan when assigned an area of operations that includes built-up areas. The British Army's **Brigadier Jonathan B. A. Bailey** was a vital player during NATO's operations to bring peace to Kosovo in 1999. Here, as in Bosnia-Herzegovina to the north, the demands on military officers to understand far more than the purely military aspects of a problem when dealing with urban operations were clearly articulated. Brigadier Bailey described the sensitive political and social concerns inherent in the peace initiative and the role played by urban areas during that process. **Colonel Gary W. Anderson** concluded the first day by summarizing the Marine Corps Warfighting Lab's ongoing efforts to improve U.S. force capabilities via their Urban Warrior and Project Metropolis experiments.

Day Two began with a division commander's view of how a force should prepare for urban operations during stability missions. **Brigadier General David L. Grange (USA, ret.)**, who commanded the U.S. 1st Infantry Division in Bosnia-Herzegovina and northwest Europe, used the lessons from his unit's Balkan deployment to better prepare his command for future contingencies in built-up areas. The general was followed by the first of seven second-day speakers who addressed strategic- to tactical-level concerns regarding operations in Somalia during the early 1990s. **Ambassador Robert B. Oakley** captivated his listeners with a recitation of the near-anarchic conditions he found in Mogadishu on first arriving and how diplomats, military authorities, and humanitarian workers combined their skills and resources in an initially successful effort to restore order to that failed nation. His strategic perspective was followed by an operational-level view provided by **Major General Carl F. Ernst (USA, ret.)** whose position as Commander, Joint Task Force Somalia, demanded skillful coordination of joint tactical actions in urban scenarios, actions that could, and eventually did, have immediate strategic consequences. Complementing these high-level analyses was **Sergeant First Class Matthew P. Eversmann's** superb recounting of actions in the streets of Mogadishu during the combat of October 3–4, 1993, which precipitated a CJTF Somalia and U.S. withdrawal from the country. SFC Eversmann was a squad leader when U.S. Ranger and Delta Force soldiers took on the missions of seizing high-ranking Somali clansmen and the rescue of American crewmen from two downed helicopters. The afternoon of the second and final day began with **Major Scott D. Campbell** recounting the many lessons learned as commander of marine snipers in downtown Mogadishu. The play of human nature, ROE, and the need to protect UN forces made Major Campbell's a most demanding task, one involving a man-weapon combination with considerable potential for addressing problems during many types of urban taskings. **LTC John Holcomb** spoke on his experiences as a surgeon in the Somali capital, experiences that included treating American casualties returning from the engagements of October 3–4. His comments on the extraordinary demands of urban operations and the need to consider alternative training methods for medical personnel were highly regarded by all in the audience. **Command Sergeant Major Michael T. Hall** and **SFC Michael T. Kennedy** followed with a recounting of how the Ranger Regiment has taken the lessons learned from Somalia and applied them to urban training for that elite unit. **Dr. James N. Miller**, Deputy Assistant Secretary of Defense for Requirements, Plans, and Counterproliferation Policy, concluded the conference with a call [on the part of the joint community] for greater efforts to improve U.S. force urban operations readiness and a proposal for a sequence of actions that would facilitate attaining that objective.

ACKNOWLEDGMENTS

A special thanks is in order for RAND's co-hosts of the 2000 conference: the men and women representing the U.S. Army Training and Doctrine Command, the United States Marine Corps Warfighting Lab, and those of the Office of the Secretary of Defense. Particularly notable in their support were Fred Dupont, Lt Col John Allison (USMC, ret.), Lt Col Larry Corbett, (USMC, ret.) and Lt Col Duane Schattle (USMC, ret.).

For the third consecutive year, the success of the conference was very much due to the extraordinary efforts of Donna Betancourt, Director of Operations, Army Research Division, RAND Arroyo Center. Two other ladies whose services to the cause were invaluable are Phyllis Switzer and Chris Hillery. Notable thanks are due to Terri Perkins, who acted as the executive manager for all conference-related events and ensured that critical tasks were never allowed to escape completion.

The entire substance of the conference was attributable to the excellent presentations of the gentlemen whose names appear in the first appendix. Without these speakers' time, effort, and willingness to take on the tough problems that continue to confront our armed services in the urban operations arena, there could have been no forum for this valuable joining of men and women interested in assisting the nation as it seeks solutions.

The timely fashion in which this document was released and the fine quality of its presentation are entirely attributable to four extraordinary women: Terri Perkins (yet again), Ann Deville, Ingred Globig, and Nikki Shacklett. Compilation of many briefings would have been impossible had it not been for the help of RAND's audiovisual gurus, Tim Lee, Marilyn Freemon, and Rodney Yeargans. The exceptional cover artwork is the work of Priscilla B. Glenn; final cover design and production is attributable to Pete Soriano and Ron Miller.

The RAND Arroyo Center's urban operations studies are very much a team effort. The editor thanks each team member for his or her help during the preparation and conduct of the affair: Lt Col Jay Bruder, Sean Edwards, Scott Gerwehr, Lois Davis, Jamie Medby, and Olya Oliker. The team thanks Dr. David Chu for graciously opening the event.

The editor also recognizes the professionalism and dedication of this document's reviewers. John Gordon of RAND and Lt Col Duane Schattle (USMC, ret.) both provided valuable observations and comments of service to our readers, as did Ambassador Robert Oakley with his review of the report's frontal material.

GLOSSARY

.50 CAL	.50 caliber machine gun
18D	Military Occupational Specialty 18D (medic)
91B	Military Occupational Specialty (MOS)—designation for a U.S. Army medic
A2C2	Army Airspace Command and Control
A6 or A6E	USN attack aircraft (Intruder)
A/Secs	Assistant Secretaries
AA	Assembly Area
AAA	Anti-aircraft artillery
AAR	After Action Review
AAV	Amphibious Assault Vehicle
ABG	Arterial Blood Gas
AC-130	Propeller-driven aircraft, models of which are used for transport, reconnaissance, and fire support by the U.S. Air Force
A/C	Aircraft
ACE	Allied Command Europe
Acft	Aircraft
ACTD	Advanced Concept Technology Demonstration
ACTORD	Activation Order

AFOR	Albania Force
AGS	Armored Gun System
AH-1F	Cobra attack helicopter, F model
AH-1W	Cobra attack helicopter, W model
AID	Agency for International Development
AIRCMD	Air Command
AJP	Australian Joint Procedures
AK-47	Model of automatic rifle manufactured predominantly by former Warsaw Pact or Communist nations
AMB	Ambassador
AMPHIBS	Amphibious Ship
AMTRAK	Amphibious Tractor
ANGLICO	Air/Naval Gunfire Liaison Company
AN-PEQ-2	Laser attachment for small arms, used for target spotting
AN-PVS 7/14	A model of night vision goggle
ANPVS-4	A model of night vision scope
AOR	Area of Responsibility
AP/AT	Anti-personnel/Anti-tank (mines)
APC	Armored Personnel Carrier
APDS-T	Armor-Piercing Discarding Sabot with Tracer
AR	Armor
ARFOR	Army Force
ARG	Amphibious Ready Group
Armd	Armored
ARRC	Allied Command Europe Rapid Reaction Corps
Art. cor.	Artillery Coordinator (forward observer)

Arty	Artillery
AS	Australian
ASAS	All Source Analysis System
AT	Anti-tank
ATGM	Anti-tank Guided Missile
ATK POS	Attack Position
ATLS	Advanced Trauma Life Saver
ATP-35 (B)	Land Force Tactical Doctrine document (U.S.)
AUST	Australian
AV-8B	U.S. Navy aircraft (Harrier)
avail	Available
AVLB	Armored Vehicular Launched Bridge
AVN	Aviation
AWACS	Airborne Warning and Control System
BD	Battle Drill
BDE or Bde	Brigade
BFV	Bradley Fighting Vehicle (M2)
Big Red One	Nickname for the U.S. Army 1st Infantry Division
BiH	Bosnia-Herzegovina
BLD	Building
BMP	Infantry Fighting Vehicle produced by several former member nations of the Warsaw Pact
BMP KSh	Command version of BMP vehicle
Bn or BN	Battalion
BOS	Battlefield Operating Systems
BP	Battle Position
br	Brigade

BREM (or BREM-1)	A tracked armored recovery vehicle (Warsaw Pact)
BTM	A high speed trenching machine manufactured by former Warsaw Pact nations
BTR	Former Warsaw Pact manufactured wheeled armored personnel carrier
Btry	Battery (Artillery unit of company size)
C	Centigrade
CA	Civil Affairs
C&C	Command and Control
C-141	A U.S. Air Force jet transport aircraft
C/J3	Staff section on combined ("C") and joint ("J") staff responsible for operations, plans, and training
C2	Command and Control
C3	Command, Control, and Communications
C3I	Command, Control, Communications, and Intelligence
CA	Civil Affairs or California
CAPT	Captain
CAS	Chief Air Staff, Chief of Air Services, or Close Air Support
CASEX	Close Air Support Exercise
CAV	Cavalry
CBT	Combat
CCATT	Critical Care Aeromedical Transport Team
CDF	Chief Defence Force
CENTCOM	Central Command

CEOI	Communications-Electronics Operating Instructions
CFS	Commander Fleet Services
CGS	Chief of the General Staff or Chief of Ground Services
CH-46	Sea Knight helicopter
CH-53	Super Stallion helicopter
CI	Counterintelligence
CINC	Commander-in-Chief
CINCCENT	Commander-in-Chief, Central Command
CISE	CENTCOM Intelligence Support Element
CIT	Counter Intelligence
CIVPOL	Civilian Police
Class I	Subsistence items, e.g., food
Class IV	Construction materials
CLFX	Combined Live Fire Exercise
CLNC	Camp Lejeune, North Carolina
CMO	Civil-Military Operations
CMOC	Civil-Military Operations Center
CMTC	Combat Maneuver Training Center
Cmte	Committee
CNN	Cable News Network
CNS	Chief of Naval Services
Cntl	Control
CO	Company
COAX	Coaxial
COCOM	Combatant Command (a command authority)

COL	Colonel
COM	Commander
COMARFORSOM	Commander Marine Forces Somalia
COMD	Command
COMINT	Communications Intelligence
COMJTF-SOM	Commander Joint Task Force Somalia
COMKFOR	Commander, Kosovo Force
COMMEX	Communications Exercise
COMMO	Communications
CONPLAN	Contingency Plan
cont	Continued
CP	Command Post
Cpl	Corporal
CPT	Captain
CQC	Close Quarters Combat
CQM	Close Quarters Marksmanship
CS	o-chlorobenzalmalononitrile, a riot control agent
CSAR	Combat Search and Rescue
CSH	Combat Support Hospital
CSS	Combat Service Support
CST	Coalition Support Team (a U.S. special forces team positioned with UN units)
CT	Computed Tomography imaging, also known as "CAT scanning" (Computed Axial Tomography)
CTF	Combined Task Force
CVGB	Carrier Battle Group
D-Day	In conjunction with Brigadier Bailey's briefing, D-Day was the day that KFOR entered Kosovo

D-1	One day before D-Day
DART	Disaster Assistance Response Team
DC	Washington, D.C. or Displaced Civilians
DCG	Deputy Commanding General
DCSOPS	Deputy Chief of Staff for Operations
DEC	December
DEMO	Demonstration
Desert Fox	Post-Persian Gulf War air operation over northern Iraq
DIC	Desiminated Intravascular Coagulation
DIV	Division
DM	Designated Marksman
DOC	Doctor
DOD (or DoD)	Department of Defense
DOS	Days of Supply
DOTMLPF	Doctrine, Organization, Training, Materiel, Leadership, Personnel, Facilities
DOW	Died of Wounds
DP	Displaced Person
Dragon	U.S. medium anti-tank weapon system
DS	Direct Support
E2C	Hawkeye tactical warning and control aircraft (USN)
EA	Each
EGS	“Everything Goes to Shit” (slang)
EiF	Entry into Force
ELINT	Electronics Intelligence
EM	East Timor

EMT	Emergency Medical Technician
EMT-P/SOMTC	Emergency Medical Technician-Paramedic/ Special Operations Medical Trauma Course
EN	Engineer
Eucaliptus	Vehicle radio with 20–50 km range (also called Romashka)
EVAC	Evacuation
ExCom	Executive Committee
F-2C	Airborne C3 platform (Banshee aircraft)
FA-6B	Electronic countermeasures/jamming aircraft
F-14A	Tomcat fighter/attack aircraft
FA	Field Artillery
FA-18C	Hornet fighter/attack aircraft
FAC	Forward Air Controller
FAC(A)	Forward Air Controller (Airborne)
FAE	Fuel Air Explosive
FAPSI	Federal Government Communications and Information Agency
FARs	Flat-Assed Rules (slang)
FAST	Forward Area Security Team or Fleet Antiterrorism Security Team
FAV	Fast Attack Vehicle (USMC)
FDA	Food and Drug Administration
FE	Force Element
FIST	Fire Support Team
FM	Field Manual or Frequency Modulation
FRY	Federal Republic of Yugoslavia

FSB	Federal Security Service (Russian) or Forward Support Battalion (U.S.)
FSE	Fire Support Element
FSK	Federal Counterintelligence Service, the forerunner of the FSB
FST	Forward Surgical Team
Ft.	Fort
FVIIa	A clotting factor
FWFE	Fixed Wing Force Element
FYROM	Republic of Macedonia
GAO	Government Accounting Office
Gen (or GEN)	General
GFAP	General Framework Agreement for Peace
GLID	Ground/Vehicular Laser Locator Designator
GOSSIP	Ground Observation Special Support Intelligence Program
GP	Group
GSM	Ground Station
GSW	Gunshot Wound
GTA	Grafenwohr Training Center (in Germany)
H&P	History and Physical examination
H-Bar	The heavy barrel version of the M-16 rifle
HA	Humanitarian Assistance
HE	High Explosive
HIV	Human Immunodeficiency Virus
HLZ	Helicopter Landing Zone
HM2	A Navy Corpsman (medic)
HMMWV	High-mobility, Multi-purpose Wheeled Vehicle

HQ	Headquarters
HRS	Human Relief Sectors
HSD	Hypertonic Saline Dextan
HUMINT	Human Intelligence
HUMVEE	Popular term for HMMWV (HUMVEE or HUMMER is also the civilian/commercial designation for the vehicle)
HVO	Croatian Defense Force
IAW	In Accordance With
IC	International Community
ICRC	International Committee of the Red Cross
ICU	Intensive Care Unit
ID	Identification
IEW	Intelligence and Electronic Warfare
IFOR	Implementation Force (during Operation Joint Endeavour/Endeavor)
IFV	Infantry Fighting Vehicle
I MEF	First Marine Expeditionary Force
IMINT	Imagery Intelligence
IN	Infantry or Inches
indep	Independent
INMARSAT	International Maritime Satellite
INTERFET	International Force East Timor
int'l	International
IO	International Organizations or Information Operations
IOT	In Order To
IPB	Intelligence Preparation of the Battlefield

IPTF	International Police Task Force
IR	Infrared
ISR	Intelligence, Surveillance, and Reconnaissance
J/C	Joint/Combined
J1	Personnel directorate of a joint staff
J2	Intelligence directorate of a joint staff
J3	Operations directorate of a joint staff
J4	Logistics directorate of a joint staff
J5	Plans directorate of a joint staff
J6	Command, Control, Communications, and Computer Systems directorate of a joint staff
JAG	Judge Advocate General
JAWP	Joint Advanced Warfighting Program
JCAHO	A civilian hospital review organization
JCATS	Joint Conflict and Tactical Simulation
JCO	Joint Commission Officer
JCS	Joint Chiefs of Staff
JFCOM	Joint Forces Command
JIB	Joint Information Bureau
JIC	Joint Implementation Commission
JMAP	Joint Military Appreciation Process
JMC	Joint Military Commission
JNLWD	Joint Non-lethal Weapons Directorate
JOC	Joint Operations Center
JOSE	Joint Support Element
JP	Joint Publication
JROC	Joint Requirements Oversight Council

JRTC	Joint Readiness Training Center (at Fort Polk, Louisiana)
JSEAD	Joint Suppression of Enemy Air Defenses
JSOTF	Joint Special Operations Task Force
JSTARS	Joint Surveillance Target Attack Radar System
JTF	Joint Task Force
JTFSOM	Joint Task Force Somalia
K	kilometer/kilometers
KC-135	Aerial refueling aircraft
K-Day	The day marking the cessation of hostilities in Kosovo. Also the day that the UCK undertaking came into force
KFOR	Kosovo Force
KHAT	A chewed stimulant, either the young buds or fresh leaves of the shrub catha edulis (also spelled kat, qat, chot, gat)
KIA	Killed in Action
KLA	Kosovo Liberation Army
km	kilometer/kilometers
KONKURS	AT-5, Spandrel (an anti-tank weapon system)
KPC	Kosovo Protection Corps
KPS	Kosovo Police Service
KTC	Kosovo Training Corps
Kub	A type of missile (Russian)
KVM	Kosovo Verification Mission
LA	Los Angeles
LAV	Light Armored Vehicle
LB	Pound (measure of weight)

LCAC	Landing Craft, Air-cushioned
LCU	Landing Craft, Utility
LDRS	Leaders
LOC	Line of Communications
LOGSUPCOM	Logistics Support Command
LRRP	Long Range Reconnaissance Patrol
LSC	Logistics Support Command
LTC	Lieutenant Colonel
LTD	Laser Target Designator
LTG (or LtGen)	Lieutenant General
LVTP (or LVT-P)	Landing Vehicle Tracked, Personnel variant; a previous nomenclature for Amphibious Assault Vehicles (AAV)
LWD	Land Warfare Doctrine
LWP	Land Warfare Pamphlet
LZ	Landing Zone
m	meter/meters
M1	U.S. manufactured Abrams main battle tank, earliest model (105-mm main gun)
M1A1	U.S. manufactured Abrams main battle tank, model A1 (with 120-mm main gun)
M1A1C	A1 model Abrams tank with improved armor and power train
M2	U.S. manufactured Bradley infantry fighting vehicle
M2A2	U.S. manufactured Bradley infantry fighting vehicle, A2 model
M3A1	Scout version of the Bradley vehicle
M9	Model of pistol

M9 ACE	U.S. manufactured combat engineer vehicle
M4	Carbine variant of the M16 rifle
M14	Standard issue rifle for U.S. Army and U.S. Marine Corps prior to the M16. Still used in sniper and other roles
M16	Standard issue rifle for USA and USMC
M16A2	Model of the M16 rifle
M21	Model of sniper rifle
M24	Model of sniper rifle
M40	Model of sniper rifle
M49	Type of spotting scope
M68	A scope used with either the M4 or M16
M106	4.2 inch (120-mm) mortar vehicle (M113 chassis)
M109 SP	Self-propelled 155-mm howitzer
M113	An armored personnel carrier (U.S. design and manufacture)
M203	Weapon system with 40-mm grenade launcher mounted below barrel of an M16 rifle
M240	7.62-mm machine gun originally coaxially mounted in M1 series tanks; being developed as infantry weapon
M249	5.56 machine gun, Squad Automatic Weapon System (SAWS)
M981 FIST V	Fire Support Team Vehicle (on an M113 chassis)
MAP	Military Appreciation Process
MARCOM	Marine Command
MARFOR	Marine Force
MAST	Military Assistance to Safety and Traffic
MC	Medical Corps

MChS	Ministry of Extraordinary Situations (akin to the U.S. Federal Emergency Management Agency, FEMA)
MCIA	Marine Corps Intelligence Activity
MCOFT	Mobile Conduct of Fire Trainer
MCWL	Marine Corps Warfighting Lab
MD	Medical Doctor
Med	Medical
MEDEVAC	Medical Evacuation
MEF	Marine Expeditionary Force
METL	Mission Essential Task List
METT-T	A decision analysis aid that considers Mission, Enemy, Terrain, Troops Available, and Time
MEU	Marine Expeditionary Unit
MFG	Manufactured
MG	Machine Gun or Major General
MGEN	Major General
MICLIC	Mine Clearing Line Charge
MIL MED	*Military Medicine*, a journal
MK-19	40-mm grenade machine gun
mm	millimeter
MNB	Multinational Brigade
MND	Multinational Division
MNF	Multinational Force
MOD or MoD	Ministry of Defense
MOOTW	Military Operations Other Than War
MOUT	Military Operations on Urbanized Terrain
MP	Military Police

mph	miles per hour
MR Co	Motor Rifle Company
MRB	Motor Rifle Brigade
MRD	Motor Rifle Division
MRMC	Medical Research and Materiel Command
MRP	Motor Rifle Platoon
MRR	Motor Rifle Regiment
MSE	Mobil Subscriber Equipment communications system
Msn or MSN	Mission
MSR	Main Supply Route
MSRT	Mobile Subscriber Radio Terminal
MTA	Military Technical Agreement
MTF	Military Treatment Facility
MTOE	Modified Table of Organization and Equipment
MTT	Mobile Training Team
MUP	Yugoslav Ministry of the Interior
MVD	Theater level command
N-Hour	Notification Hour, the time a unit receives notification of an alert
NATO	North Atlantic Treaty Organization
NAV	Navy or Naval
NAVCENT	U.S. Naval Forces Central Command
NBC	Nuclear, Biological, Chemical
NC	North Carolina
NCA	National Command Authorities
NCO	Noncommissioned Officer

NCOPD	Noncommissioned Officer Professional Development
NEO	Noncombatant Evacuation Operation
NGO	Nongovernmental Organization
NJ	New Jersey
NL	Nonlethal
NMCC	National Military Command Center
No	Number
NODS	Night Observation Devices
NORD	North (NORD Brigade in Bosnia consisted of Scandinavian and Polish elements)
NOTAM	Notice to Airmen
NSC	National Security Council
NTC	National Training Center (at Fort Irwin, California)
NVD	Night Vision Devices
NW	Northwest
NY	New York
NZ	New Zealand
NZDF	New Zealand Defence Force
NZFOREM	New Zealand Force East Timor
O/H	On Hand
OH-58A	Kiowa light observation helicopter
OH-58H	Kiowa Warrior reconnaissance helicopter
OIC	Officer in Charge
OMS	Generic designation for a separate motorized rifle brigade (Russian)
OOTW	Operations Other Than War

OP	Observation Point; or designation for an operational level task from the Universal Naval Task List
OPCOM	Operational Command
OPCON	Operational Control
OPD	Officer Professional Development
OPLAN	Operations Plan
OPORD	Operations Order
Ops	Operations
OPSEC	Operations Security
OPTEMPO	Operating Tempo
OR	Operating Room
ORJ	Orasje
OSD	Office of the Secretary of Defense
OSI	Office of Special Investigations
PA	Physician's Assistant
PAO	Public Affairs Officer
para	parachute
PB	Parachute Brigade
PD	Parachute Division
PDD	Presidential Decision Directive
PFC	Private First Class
PK	A type of machine gun (Russian or former Warsaw Pact manufacture)
PKF	Peacekeeping Force
PKM	5.45-mm machine gun (Russian or former Warsaw Pact manufacture)
Plt	Platoon

Plt Cmdr	Platoon Commander
Plt RTO	Platoon Radiotelephone Operator
Plt Sgt	Platoon Sergeant
PM	Provost Marshal
PMI	Pre-marksmanship Instruction
PMO	Provost Marshal Office
POD	A pod attached to aircraft for photo reconnaissance
POI	Program of Instruction
POS	Posvina
POTF	Psychological Operations Task Force
PPBS	Planning, Programming, Budgeting System
PR	Pilot Recovery
PSO	Peace Support Operations
PSYOP	Psychological Operations
PT	Physical Training
PT-76	A lightly armored amphibious tank, Warsaw Pact
PVO	Private Voluntary Organization
PWG	Posovina Working Group
Q36	Model number for a counter-fire radar
QRF	Quick Reaction Force
RAS	Recovery Activation System
R&D	Research and Development
R&S	Reconnaissance and Surveillance
R-145 KB	A shortwave radio
RACK	Personal load-bearing system for carrying equipment on the body

RADM	Rear Admiral
RECCE	Reconnaissance
REDCON	Readiness Condition
REEF POINT	U.S. Navy intelligence collection system
REMBASS	Remote Battlefield Sensor System
Reps	Representatives
ret	Retired
RFCT	Ready First Combat Team
RNZAF	Royal New Zealand Air Force
ROE	Rules of Engagement
Romashka	Vehicle radio with 20–50 km range (also called Eucaliptus) (Russian)
RPG	Rocket-Propelled Grenade (also used in reference to launcher)
RPK-74	Machine gun version of AK-74 5.45-mm rifle (Russian)
RPO	Designation for thermobaric weapons such as the RPO-A “Shmel”
RS	Serb Republic of Bosnia
RSM	Regimental Sergeant Major
RSTA	Reconnaissance, Surveillance, and Target Acquisition
RT	Response Time or Respiratory Technician
RTC 350-1-2	Ranger Training Circular 350-1-2, a manual
RTO	Radio Telephone Operator
RWFE	Rotary Wing Force Element
S2	Battalion, regimental, or brigade intelligence staff officer
S-3B	Viking submarine detection and attack aircraft

S&T	Science and Technology
SA8	Gecko (Osa)
SAAR	Somalia After Action Review
SAM	Surface-to-Air Missile
SAMS	School of Advanced Military Studies
SAR	Search and Rescue
SAS	Special Air Service
SASO	Stability and Support Operations
SCR	Security Council Resolution
SD	Storm Detachment
SEAD	Suppression of Enemy Air Defenses
SEALs	Sea-Air-Land Teams
SEE	Small Equipment Excavator, U.S. Army engineer vehicle
SEM	Semberija, a town in Bosnia-Herzegovina
SF	Special Forces
SFC	Sergeant First Class
SFOR	Stabilization Force
Sgt	Sergeant
SH-3H	Sea King helicopter
Shilka	ZSU-23/4 anti-aircraft weapon system
Shmel	Shoulder-fired thermobaric weapon
SIGINT	Signals Intelligence
SIM	Simmunitions (a brand of frangible training ammunition)
SIMRAD	Type of night vision scope
SJA	Staff Judge Advocate

SNA	Somali National Alliance
SNCO	Staff Noncommissioned Officer
SNO	Senior National Officer
SOA	Special Operations Aviation
SOCOM	Special Operations Command
SOCCE	Special Operations Command and Control Element
SOF	Special Operations Forces
SOI	Signal Operating Instructions
SOM	Somalia
SOP	Standing Operating Procedures
SOW	Special Operations Wing
SP	Self-propelled
SQD	Squad
Sqds	Squads
SR-25	7.62-mm rifle
SRBK	Srebnik, a town in Bosnia-Herzegovina located north of Tuzla
SRSG	Special Representative Secretary General
SSgt	Staff Sergeant
STX	Situational Training Exercise or Squad Tactical Exercise
SUPCOM	Supply Command
SVD	A sniper rifle (Russian)
SWSS	Secure Weapon Storage Site
T	Trained
T-72	Tank model that was produced by several former member nations of the Warsaw Pact

TAC or TAC CP	Tactical Command Post
TACAIR	Tactical Air support
TACON	Tactical Control
TACSAT	Tactical Satellite
TARPS	Tactical Airborne Reconnaissance Pods System
TB	Tuberculosis
TC	Track Commander
TCP	Temporary checkpoints or traffic control points
TEWT	Tactical Exercise Without Troops
TF	Task Force
TGT	Target
Tier 3 extraction	A contingency plan to secure the withdrawal of the KVM from Kosovo
TIO	Target Information Officer
TM	Team
TNI	Indonesian Military Forces
T/O	Task Organization
TOS-1	220-mm multiple rocket launcher, a Russian weapon system
TOW	Tube-launched, Optically-tracked, Wire-guided missile (U.S.)
TPQ-36	AN/TPQ-36 fire finder radar. Also referred to as Q-36
TRAP	Tactical Recovery of Aircraft and Personnel
TTP or TT&P	Tactics, Training, and Procedures
Tunguska	2S6, an anti-aircraft weapon system
TV	Television
TX	Texas

U/S	Under Secretary
UAV	Unmanned Aerial Vehicle
UCCATS	Urban Combat Computer Assisted Training System
UCK	Ushtria Clirimtare e Kosoves (Designation of the KLA in Albanian)
UH1H	Model of transport helicopter (U.S. design and manufacture), H model
UH-1N	Model of transport helicopter (U.S. design and manufacture), N model
UH-1V	Model of transport helicopter (U.S. design and manufacture), V model
UH-60L	U.S. Black Hawk utility helicopter, L model
UK	United Kingdom
UN	United Nations
UNAMET	United Nations Assistance Mission in East Timor
UNICEF	United Nations Children's Fund
UNISOM II	United Nations in Somalia II
UNITAF	Unified Task Force
UNMIK	UN Mission in Kosovo
UNNY	United Nations New York
UNOSOM	United Nations Operation, Somalia
UNPROFOR	United Nations Protection Force
UNSC	United Nations Security Council
UNSCR	United Nations Security Council Resolution
UNSRSG	United Nations Special Representative of the Secretary General
UNTAET	United Nations Transitional Administration in East Timor

UNTL	Universal Naval Task List
UR-77	An explosive hose mine clearing system mounted on a 2S1 chassis (Russian)
U.S.	United States
USA	United States Army or United States of America
USAF	United States Air Force
USAREUR	United States Army, Europe
USCG	United States Coast Guard
USCINCCENT	United States Commander in Chief Central Command
USFORSOM	United States Forces Somalia
USLO	United States Liaison Officer
USMC	United States Marine Corps
USN	United States Navy
USS	United States Ship
USSOCOM	United States Special Operations Command
UTD	Up to date
UWG	Urban Working Group
V8 LR	V8 Land Rover (a wheeled vehicle)
VDD	Generic designation for an airborne division (Russian)
VIP	Very Important Person
VJ	Yugoslav Army
Vol	Volume
VRS	Bosnian Serb Army
VS	versus
VTOL	Vertical Take-off and Landing
WANS	Wide Area Networks

WARNORD	Warning Order
WIA	Wounded in Action
WMD	Weapons of Mass Destruction
WPNS	Weapons
WSS	Weapons Storage Site
WWI	World War I
WWII	World War II
X	Military symbol for a brigade
XO	Executive Officer
XX	Military symbol for a division
XXX	Military symbol for a corps (e.g., II XXX would represent II Corps)
ZOS	Zone of Separation
ZSU-23-4	Shilka anti-aircraft gun system, Warsaw Pact (also referred to as ZSU-23 in briefings)

INTRODUCTION

> **Capital.** A capital town or city; the head town of a country, province, or state.
>
> **Capital.** The accumulated wealth of an individual, company, or community, used as a fund for carrying on fresh production;wealth in any form used to help in producing more wealth.
>
> —*The Oxford English Dictionary*

These two meanings of a single word ironically underlie the reasons urban areas play an increasingly significant role in many of the world's recent force deployments. Cities, indeed *capital* cities in particular, more often than not have had a primary role in the most recent of the United States' major military operations. Seoul exchanged hands several times during the Korean War of 1950–1953. Both the ancient capital of Hue and the modern capital of Saigon played strategically significant, some would argue decisive, roles in the eventual withdrawal of the United States from Vietnam after Tet 1968. The loss of 241 men in the 1983 bombing of the Marine barracks in Beirut similarly led to the removal of U.S. forces from Lebanon and a failure of U.S. policy. Operations during the 1989 Operation Just Cause centered on locations in and around Panama City. Just over a year later, the retaking of Kuwait City symbolized the liberation of the nation for which that metropolitan area serves as capital. It was Mogadishu that held the attention of U.S. and UN forces in 1992–1993, and it was the loss of two helicopters and eighteen soldiers there in October 1993 that precipitated the U.S.

removal of its forces in the months immediately following those events. Beirut, Monrovia, Freetown, Belgrade, Sarajevo, and Port-au-Prince saw America's soldiers and marines in their streets or its pilots overhead during this same period.

Few should be surprised that this is the case. Cities, capitals especially, are focal points for much of the capital wealth and human resources of modern nation states. Commercial enterprise aggregates where resources coalesce, and it is to cities that much of a nation's raw material comes to be formed into finished products, capital ready for employment in generating yet further wealth for citizen and community. Those who live in and around these urban areas in increasing numbers both create and consume the products born of these capital goods. They comprise an often combustible social capital, the role of which is much like the life's blood of a nation. The people flow through city streets as does blood through arteries and veins; without them the city is as inert and lifeless as a bloodless body. This social capital fosters yet other forms of wealth—emotional, political, cultural, and economic—that themselves bolster or hamper urban residents in their ongoing efforts to promote the growth of individual and community health. Modern cities are the seats of the world's capital.

It is therefore not surprising that the condition of a nation's cities is often a reflection of that country's welfare. Anarchy in Sarajevo signaled the collapse of order in Bosnia-Herzegovina. The ruins of Grozny reflect the loss of Chechnya as a cooperative member of the Russian state. Mogadishu's dissolution into clan struggles reflected the absence of a governing body able to care for the nation's people. The destruction of the social and physical capital of cities, whether due to internal causes or external incursion, marks the decline of a nation as a viable member of the world order. Armed forces that ignore this relationship may succeed in accomplishing military tasks only to fail in serving political objectives. Yet even military success is not assured; history demonstrates that the demands of urban combat pose a vital threat to a force's capital: its manpower, equipment, and sustenance. Fighting in urban areas can precipitate extraordinary losses of men and equipment and ensuing policy failure. In many ways, then, success in the conduct of military operations means preserving capital: that of the citizens living in a city, of the infrastructure on which they depend, and of the force itself.

It was recognition of urban areas' significance in today's world that drove RAND Arroyo Center, U.S. Army Training and Doctrine Command (TRADOC), Marine Corps Warfighting Lab (MCWL), and the Office of the Secretary of Defense (OSD) to host a third annual conference to promote better understanding of how nations can conduct less costly yet successful operations in built-up areas. The briefings spanned the spectrum from support missions in Mogadishu to stability actions in the towns of East Timor to combat in Grozny. Their relevance encompassed activities at the tactical level of individual snipers to strategic concerns briefed by a former U.S. ambassador to Somalia and Russian military commander in Chechnya. Virtually every type of military concern was evident: medical care for wounded Somalis and friendly forces; aircraft recovery; chemical weapons use; riot control; nation building; force protection; PSYOP; organized crime; technological innovation; civil affairs; joint, multinational, and combined arms operations; training; political policy; acquisition; and logistical support are but a sample.

Urban Operations and the Modern Strategic Environment

As was the case during the 1998 and 1999 conferences, this event spawned difficult questions that challenged both speakers and listeners. How, for example, does a military prepare for habitually complex and heterogeneous urban operations? What lessons can be gained from examples as diverse as Russian forces fighting with minimally restrictive rules of engagement (ROE) in Grozny and highly constrained U.S. actions during stability missions in Brcko, Bosnia-Herzegovina? Together, presenters and audience made strides toward better understanding such challenges and in determining possible solutions to these and other questions that were raised over the two-day period.

The briefings contain many good points from which the reader can draw valuable lessons. Three such lessons have particularly wide-ranging application. *First,* a similar observation from presentations on very different types of urban operations reflects that cities are no longer purely a military concern. Ambassador Oakley, Generals Kulikov and Grange, and Colonel Fontenot all noted the tight linkage between tactical activities and strategy during their respective operations in Somalia, Chechnya, and Bosnia-Herzegovina. In all instances this linkage was a function of more than military actions alone. These leaders cited the need to synchronize elements of both civilian and military power and influence; any one element alone possessed insufficient scope to meet the combined demands of national, coalition, or private organizational interests. Unilateral efforts by military, diplomatic, or other representatives can in fact prove counterproductive; synchronization during urban contingencies demands more than superficial cooperation or sharing a headquarters. Participation in goal definition, plan development, and operational control must be continuous and mutually supporting.

That the Cold War's two greatest superpowers have each suffered strategic defeats in urban areas at the hands of loosely organized, armed bands within the past decade offers a *second lesson* of particular note. In the traditional World War II sense, it is arguable that Mogadishu and Grozny were victories: the Americans conducted a successful raid to capture enemy personnel; the Russians twice took the capital of Chechnya. Yet a half-century in time has altered perceptions of victory. The loss of eighteen U.S. soldiers was perceived by decisionmakers in Washington as an unacceptable cost when the goals sought were seen as peripheral to American interests. The capture of Grozny was perceived to be poor compensation when the price was the deaths of hundreds of Russian conscripts.

The *third* primary lesson is evident in considering the causes that underlie many recent military deployments. Armed forces have customarily been perceived as nations' coercive tools of influence. Today they are increasingly also seen as agents of social improvement. Today's militaries not only defeat the wayward dictator's rogue armies; they bring relief to the suffering and protect innocents whose welfare is threatened. Commitments such as those to East Timor, Bosnia, Kosovo, and Somalia are less likely to be justified with arguments citing national interests than others touting supranational humanitarian goals. A nation's citizens react with revulsion when they look at their television screens and see their soldiers destroying civilian infrastructure or killing noncombatants. Occupation of a city rendered uninhabitable precipitates scorn rather than celebration. A campaign plan that defines success as urban terrain seized without cognizance of today's domestic and international public opinion is an ill-conceived one. Such plans must look beyond the immediate military objective to the post-combat state of affairs and coordinate the two. Victory today is less defined by raising a flag over the capitol than by rapidly returning the battleground to a semblance of preconflict normalcy. In 1945 the United States destroyed the city of Manila in the process of retaking it from the Japanese, and promptly declared the operation a success. It is highly unlikely we could do so in today's environment.

Observations

From the above-noted lessons come several observations regarding how future urban operations might be better undertaken. First, there is a call for bringing all relevant actors together in the service of strategic objectives. For example, though the goals of the UN, U.S. Army, Doctors Without Borders, the British Foreign Ministry, and CNN are highly unlikely to dovetail perfectly during an international undertaking, it would be unusual were they not to have considerable overlap. Just as coalitions of armed forces have become the typical *modus operandi* for military operations, so coalitions of relevant civilian and military organizations could emerge as the logical means of addressing the multifaceted demands of urban operations. Such cooperative ventures would combine formal authority relationships with voluntary pseudo-partnerships allowing both effective management and retention of the functional independence that NGOs and PVOs tend to demand. A closer relationship between military and diplomatic representatives in 1993 Mogadishu would most likely have precluded the U.S. military's attack helicopter strike on moderate members of the Habir Gidr clan who were meeting in an attempt to peacefully resolve the conflict between Mohammed Aideed and UN forces, an attack that led to the costly Somali battle with Task Force Ranger on October 3–4. Though international relief organizations often feel it essential to preserve their independence from government organizations, informal agreements offering security and logistical support in return for cooperation during humanitarian assistance operations should be workable in many instances. Independent bureaucratic agendas hamstrung U.S. operations in Vietnam and Mogadishu. No military force, regardless of size, can unilaterally fully comprehend, much less handle, all the problems that will arise during a major urban operation in today's world. Both immediate and longer-term success demand a systematic and well-planned joining of governmental, PVO, NGO, and commercial capabilities in the service of multinational and interagency goals.

Second, the concepts of victory and success in urban areas have assumed new forms. It is highly unlikely that western publics or their political leaders would accept another Dresden or Aachen. Armed forces habitually use a backward planning process: their representatives envision the desired end state and plan backward in time from that point, determining the steps essential to the end state's accomplishment. This end state can no longer be a flattened city with thousands of civilians picking through debris for scraps of food; it should instead be a state of relative normalcy in which the city remains a livable environment capable of sustaining the lives of its residents. Accepting this view as the objective would have obvious effects on the conduct of military operations in a city.

Third, military and other coalition members must understand that the media are part of their environment. From a purely military perspective, the loss of two helicopters on October 3, 1993, even the loss of eighteen soldiers killed, were tactical events of significant but limited consequence. Only the immediate reporting of the episode, complete with footage showing the body of an American crew member being dragged through the streets, elevated it to strategic consequence. Coalition members recognize that a media consisting of both objective and biased reporters constantly monitors their activities. It is essential that coalition members consider not only the *intentions* underlying their actions, but also the *appearance* those actions project. They must also be aware of the strategic implications of seemingly minor events. Demands by one coalition member to commit a force to a routine security mission during food distribution could compromise the entire international operation if the security force is taken hostage or ambushed and its nation embarrassed. Attaining such cognizance means that the various coalition members will have to participate in common planning sessions, sessions that include the identification of each member's concerns and the "wargaming" of various possible outcomes of given courses of action.

Fourth, as the battles of Grozny and Mogadishu tell us, it is too easy for a sophisticated, technologically advanced military force to badly—perhaps fatally—underestimate the capabilities and will to fight of a seemingly rag-tag opposing force operating in an urban environment. In both cases, the unexpectedly severe losses had not only a profound military effect but also a politically decisive one.

The following briefings stimulate other observations with immediate implications for strategic and operational planning. In his discussion of urban riots in Bosnia-Herzegovina, Colonel Jim Greer observed that tracked infantry fighting vehicles (IFVs) were far superior to wheeled systems in meeting the unpredictable conditions confronted during those actions. The IFV's neutral steer capability (the ability to rotate the tracks on each side of the vehicle in different directions so as to turn it around within a single vehicle length) allowed rapid movement out of a restricted area and intimidated hostile crowds. The three-point turns required by today's wheeled vehicles, on the other hand, put those vehicles' occupants at risk. The implications for force design both in the immediate and longer terms are clear.

Ultimately, however, none of the historical lessons learned or observations about imminently available technologies offered solutions that would dramatically reduce friendly force or noncombatant losses should a U.S. force have to engage in urban combat. There remains a desperate requirement to develop means of accomplishing urban combat missions that do not unacceptably diminish the human capital of U.S. armed services, those of its allies, or that of the cities in which they will have to fight.

Appendix A

CONFERENCE AGENDA

Wednesday, 22 March

0800–0830 **Registration and Continental Breakfast**

0830–0845 **Welcome and Introduction**
Dr. David S. C. Chu
Vice President, Army Research Division and
Director, RAND Arroyo Center

0845–0945 **The First Battle of Grozny**
General Anatoly Sergeevich Kulikov
Russian Ministry of Internal Affairs

0945–1030 **The Two Sides of Grozny**
Mr. Arthur L. Speyer, III
Marine Corps Intelligence Activity

1030–1050 **Break**

1050–1135 **The Urban Area During Stability Missions—**
Case Study: East Timor
LtCol Phil Gibbons
G3 Land Command, New Zealand Army

1135–1220 **Force Protection During Urban Operations—**
Case Study: Mogadishu (Part 1)
LtCol John Allison
U.S. Marine Corps (retired)

1220–1320 **Lunch**

1320–1405 **The Urban Area During Stability Missions—**
Case Study: Bosnia–Herzegovina
COL Greg Fontenot
U.S. Army (retired)

1405–1450 COL James K. Greer
U.S. Army

1450–1510 **Break**

1510–1555 **The Urban Area During Stability Missions—**
The British Experience in Kosovo
Brigadier Jonathan Bailey
British Army

1555–1625 **Applying the Lessons Learned—**
Take 1 (Project Metropolis)
COL Gary Anderson
U.S. Marine Corps, Marine Corps Warfighting Lab

1625–1630 **Closing Remarks for Day 1**

1630–1730 **Wine and Cheese Social**

Thursday, 23 March

0800–0830 **Continental Breakfast**

0830–0915 **Training for Urban Operations**
MG David L. Grange,
U.S. Army (retired)

0915–1000 **The Urban Area During Support Missions—**
Case Study: Mogadishu (Part 2)
The Strategic Level
Ambassador Robert B. Oakley,
former ambassador to Somalia

1000–1020 **Break**

The Urban Area During Support Missions—
Case Study: Mogadishu (Part 3)

1020–1105 **The Operational Level**
MG Carl F. Ernst,
U.S. Army (retired)

1105–1150 **The Tactical Level I**
SFC Matthew Eversmann,
U.S. Army

1150–1250 **Lunch**

The Urban Area During Support Missions—
Case Study: Mogadishu (Part 4)

1250–1335 **The Tactical Level II: The Offensive and Defensive Use of Urban Snipers**
MAJ Scott D. Campbell,
U.S. Marine Corps

1335–1420 **Medical Support**
LTC John Holcomb,
U.S. Army

1420–1440 **Break**

1440–1525 **Applying the Lessons Learned—Take 2**
CSM Mike Hall and
SFC Michael T. Kennedy
U.S. Army

1525–1610 **Concluding Remarks**
Dr. James N. Miller,
Deputy Assistant Secretary of Defense for
Requirements, Plans, and Counterproliferation Policy

1610–1615 **Closing Remarks**

Appendix B

THE FIRST BATTLE OF GROZNY
General Anatoly Sergeevich Kulikov
Russian Ministry of Internal Affairs

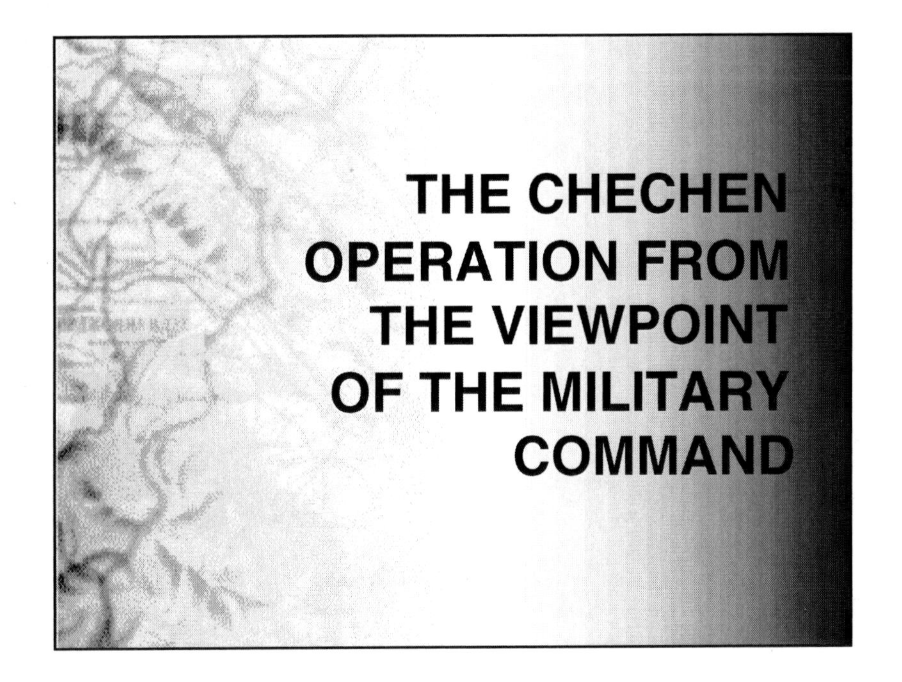

John Barry, writing in the February 21, 2000 *Newsweek*, characterized the Grozny experience as the "new urban battlefield." He quoted a senior Pentagon official as saying that he wasn't sure U.S. forces would "do a whole lot better than the Russians." Barry's article highlights the key issue increasingly recognized by analysts and planners: mortal combat in urban terrain, like that in Grozny, is combat on very unfavorable terms, and that as urbanization continues, experiences such as Grozny should be studied for their implications for the future. Barry recalls that "The U.S. Army used to have a simple way of dealing with cities: avoid them. The cost in street-fighting casualties was just too steep. That was one reason that in 1945 the Army didn't try to take Berlin, a battle that Gen. Omar Bradley told Dwight Eisenhower "might cost us 100,000 men." Bradley was right. The Red Army, which did fight its way into Berlin, lost 102,000 men doing so; 125,000 German civilians died in the battle, as did 150,000 to 200,000 German troops."[1]

---

[1] Barry's article makes a good starting point for my own presentation that will do as he suggests and consider the Grozny operations of 1994–1995 in the hope that we can learn from them.

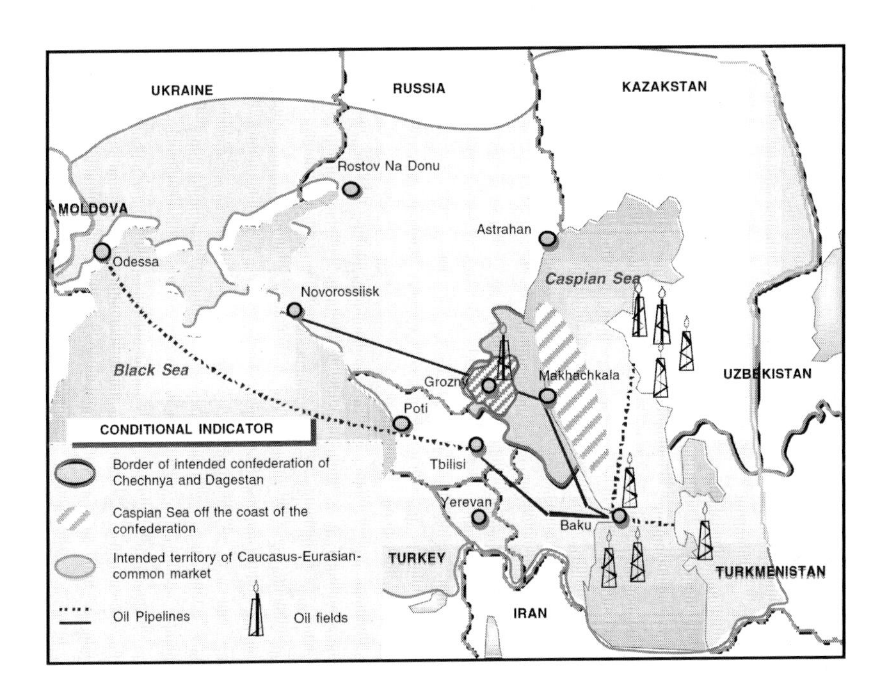

I will briefly outline the situation in Chechnya that led to the military operations of 1994–1996. Chechen separatists headed by Djohar Dudaev were seeking to create a single trans-Caucasian republic that they envisioned as stretching to include parts of Russia and the Ukraine as well as all of the Caucasian and trans-Caucasian region. This was of significant concern as the Caucasus is a strategically important area. Critical oil and natural gas pipelines traverse it, as do marine trade routes through the Black Sea to the Middle East. Thus, from the geopolitical point of view, the Caucasus is a key military and geostrategic location, a door to the Middle East.

Preconditions of the Military Operation

- Ongoing flagrant violations of the Constitution of the Russian Federation in the Chechen Republic
- Refusal of Djohar Dudaev to seek a peaceful resolution of the crisis
- Sharp increase in criminal activity
- Violations of the rights and freedoms of citizens
- Seizure and holding of hostages
- Increased deaths among the civilian population

By the spring of 1994 Dudaev had managed to turn Chechnya into the criminal center of Russia. Murders in the region were up to 2,000 people per year (in Rostov, the annual rate was 450). Fraudulent operations and schemes originating in Chechnya were destroying Russia's financial system. In May, June, and July of 1994, hostages were repeatedly seized, the perpetrators demanding helicopters and/or cash in amounts of 8–15 million dollars. As a result, the Russian speaking population was increasingly fleeing the area, unwilling and unable to continue to live under such circumstances. It was the combination of these factors that led the Russian government, or more specifically the presidential administration, to recommend to the President that steps be taken to restore constitutional order in the Republic of Chechnya. Initial negotiation efforts were made at the end of November of 1994 in conjunction with Chechen forces opposed to the Dudaev regime. Unfortunately, these proved fruitless.

Legal Basis for the Planning and Implementation of the Operation

- **Article 88 of the Constitution of the Russian Federation**
- **Decree of the President of the Russian Federation of November 30, 1994 No. 2137C: "Reestablish constitutional law and order in the territory of the Chechen Republic."**

On November 30, 1994, in accordance with Article 99 of the Constitution of the Russian Federation, the President signed Decree No. 2137c, "On steps to reestablish constitutional law and order in the territory of the Chechen Republic." The reasons for developing such steps were enumerated in the decree: the blatant violations of the Constitution of the Russian Federation in the Chechen Republic, the refusal of D. Dudaev to seek a peaceful resolution to the crisis, the increase in general criminal activity, continuing violations of the rights of and freedoms of citizens, repeated incidents of hostage-taking, and the increasing numbers of murders.

Group Composition

- Grachev, P.S. (director)
- Yegorov, H.D.
- Yerin, V.F.
- Kruglov, A.S.
- Kulikov, A.S.
- Nikolayev, A.I
- Panichev, V.N
- Pastuhov, B.N.
- Starovoitov, A.V.
- Stepashin, S.V.
- Shirshov, P.P.
- Yushenkov, S.N

In accordance with this decree, a special group was created to direct operations to disarm and eliminate illegal armed formations in Chechnya and to declare and administer a state of emergency in the territory of the Chechen Republic. The group was headed by Minister of Defense Pavel Grachev, who was also granted the rights and privileges appropriate to creating and directing the Joint Grouping of Federal Forces that was to carry out these operations.

Given the situation in the region and the impossibility of resolving the conflict by political means, an addendum to the presidential decree laid out a series of tasks to be undertaken by a number of relevant ministries and organizations and the time frames for accomplishing them.

Tasks of the Joint Grouping of Forces as Directed by the President of the Russian Federation

- **Stabilize the situation in the Chechen Republic.**
- **Disarm illegal armed bands and, in the event of resistance, destroy them.**
- **Reestablish law and order in the Chechen Republic in accordance with legislation of the Russian Federation.**

The President entrusted the Joint Grouping of Forces with the tasks shown here.

Unique Aspects of Planning for the Operation

- **A similar operatio n had never been planned and implemented previo usly.**
- **No prior decision to undertake advance planning taking into account the current condition and relative strength of forces.**
- **Combined composit ion of forces and equipment, complex system o f command of forces.**
- **No unity of publ ic opinion on the Chechen crisis.**

For all practical purposes, this was the first time such an operation had ever been planned and implemented. It was unique in the following five respects.

- First, there had been no previous decisions made nor any steps taken to plan and prepare for an operation to disarm the illegal armed bands. This was true despite the lengthy period during which the internal political situation had been developing in these unconstitutional directions even as the crisis continued to ripen and attempts by the Russian government to resolve these problems peacefully proved ineffective. In the end, the decision itself was made too late to facilitate the necessary preparation and planning.
- Second, a Joint Grouping of Forces was created to implement this operation. This grouping was composed of army units, Internal Affairs Ministry units, border guards, railroad troops, and forces and components of the FSB, FAPSI, and MChS.

- Third, a special command was created on the foundation of the existing Northern Caucasus Military Region to direct this Joint Grouping. It included a wide range of ministries and agencies.
- Fourth, never before had all of these organizations been combined in a single joint command. There was, therefore, no prior experience in coordinating such operations.
- Fifth, and finally, the operation in question was to be carried out in a component republic of the Russian Federation, creating a range of substantive constraints and operational peculiarities.

Operational Plan

With the goal of disarming illegal armed bands and confiscating weapons and armaments from the population and reestablishing constitutional law and order on the territory of the Chechen Republic, the formations and units of the armed forces, together with other military forces of the Russian Federation, are to implement a special operation in four stages.

The operational plan for the use of force in this mission was developed by the General Staff of the Armed Forces of the Russian Federation with the involvement of all other relevant ministries and agencies.

It stated the goal of the mission, as indicated, and laid out a four-stage operation to accomplish it.

Operational Plan

Stage 1 (November 29 to December 6, 1994)

- **Create force groupings for operations towards Mozdok, Vladikavkaz, and Kizliar and by December 5 take control of the starting positions for the operation.**
- **By December 1, shift tactical aviation and ground forces' aviation assets to appropriate airfields. Simultaneously establish air defense system to ensure complete defense of the airspace over the Chechen Republic.**
- **Prepare allocated electronic warfare assets to destroy the command system of illegal armed bands on the territory of the Chechen Republic.**

Stage One (November 29–December 6, 1994):

The first stage was to begin on November 26 and be completed by December 6. During this time, force groupings were to be created in preparation for troop movements in three directions: Mozdok, Vladikavkaz, and Kizliar.

By December 5, forces were to have taken control of these locations.

Before that, by December 1, tactical aviation and ground forces aviation were to have been shifted to the appropriate airfields and air defense systems were to have been in place.

At the same time, electronic warfare assets were to have been prepared in order to suppress the illegal bands' communications systems.

Operational Plan

Stage 2 (December 7–9)

- **Advance force groupings with air support advance on Grozny from six directions and blockade the city, creating an inner and outer ring.**
- **Allocate some forces to blockade larger populated areas controlled by illegal armed bands and disarm them.**
- **Internal Affairs Ministry Forces to protect communications and prevent the approach of armed groups from territory contiguous to the Chechen Republic.**
- **Special forces subunits of the FSK and Internal Affairs Ministry to ensure the isolation of individuals capable of leading diversionary operations and armed actions in the rear.**

Stage Two (December 7–9):

With air support, the advance force groupings were to advance on Grozny from six directions and blockade it by forming two concentric rings. The outer ring was to coincide with the administrative border of Chechnya and the inner ring with the outside limits of the city of Grozny.

The main part of the Russian force was to be allocated to the blockade mission and to disarming armed bands located within the city. A smaller portion of the force was to blockade other populated areas controlled by illegal armed bands and disarm those bands.

Internal Affairs Ministry Forces were to protect communications, prevent attacks on federal forces by the illegal armed bands, and prevent any approach of armed groups and detachments from territory contiguous to the Chechen Republic.

Internal Affairs Ministry and FSB special forces groups were to locate and isolate the leadership of republic, local government, and opposition parties that might be capable of leading military operations and diversions in the rear.

Operational Plan

Stage 3 (December 10–13)

- **Formations and units advance from the north and south to capture the presidential palace, government buildings, television and radio facilities, and other important structures [in Grozny].**
- **Then, together with special forces subunits of the Internal Affairs Ministry and FSB, continue to confiscate weaponry and materiel.**

Stage Three (December 10–13):

From December 10 through 13, Russian Army units were to advance from the north and south to capture the city and its key nodes (such as the palace and other government buildings). With these under control, the military was to undertake operations to disarm illegal formations.

Operational Plan

Stage 4 (December 15–23)

- **Formations and units of the armed forces to stabilize the situation and transfer zones of responsibility to the troops of the Internal Affairs Ministry who had been tasked with finding and confiscating weaponry and armaments from illegal armed bands and the population at large throughout the Chechen Republic.**

Stage Four (December 15–23):

During this stage the armed forces were to generally stabilize the situation, transferring responsibilities for critical areas to the internal troops which in turn would continue to confiscate weapons and disarm bands.

As already noted, Internal Affairs Ministry forces were tasked with creating the outer ring of control coinciding with the Chechen Republic border and with protecting the rear of the army forces. Army forces were to advance from that outer ring inwards, transferring authority to Internal Affairs Ministry forces as they captured territory and eventually entered Grozny where the procedure would be repeated. To accomplish this, the armed forces were divided into the three groups described on the following pages, each to advance on Grozny from a different direction: Vladikavkaz, Mozdok, and Kizliar. The forces in question were predominantly those of the North Caucasus Military District Army forces augmented by paratroopers and some Internal Affairs Ministry troops.

From Mozdok—Grouping 1

under the command of the First Deputy Commander of the Northern Caucasus Military Distri ct General-Lieutenant Chilindin, V.M.

- **Battalions—15** (motor rifle—2, tank—1, airborne—3, internal forces—9)
- **Battalion tactical groups of the internal forces—96**
- **Special Forces Companies of the internal forces—2**
- **SP bn—2, air defense bn—3, anti-air arty bn—1, howitzer btry—1, anti-tank btry—1**
- **Personnel—6,567**
- **Tanks—41, BTRs—99, BMPs—132**
- **Weaponry and mortars—54**

This chart outlines the forces that made up the Mozdok grouping. They included:

- From the Northern Caucasus Military District forces:
  - Forces of the 131st Independent Motor Rifle Brigade (MRB) that included about 7,000 men, 20 tanks, 50 BMPs, 4 KONKURS (AT-5 Spandrel) guided anti-tank missiles, 6 TUNGUSKA (2S6) gun-missile air defense vehicles, and other weapons systems;
  - The 481st SAM regiment of the 19th Motor Rifle Division (MRD) with 2 Osa (SA8 Gecko) SAMs;
  - A battalion from the 170th Engineer-Sapper Brigade that included sapper and road engineer companies and a pontoon bridging battalion from the 173rd Pontoon Bridging Brigade; and
  - The combined regiment of the 22nd Independent Spetsnaz brigade.

- From the Paratrooper Forces:
  - The combined forces of the 106th Paratrooper Division with 850 men in 2 airborne battalions, an SP battalion, 43 BMPs, and 39 air defense weapons; and
  - The 56th Independent Paratrooper Brigade with 3 airborne companies, an artillery and an anti-tank battery, and a total of 416 personnel with 8 units of grenade launchers and other equipment and 6 units of air defense weaponry.
- From the Internal Affairs Ministry Troops:
  - The 59th Operational Designation Regiment including two battalions, an armored group, an air defense battalion, a special forces company, 15 BTR-70s, 18 BMP-2s, 6 units of air defense weaponry, and a total of 600 personnel;
  - The 81st Operational Designation Regiment with three battalions, an armored group, a special forces company, an air defense battalion, 21 BTR-80s, 6 units of air defense weaponry, 9 PT-76s, 12 BTMs, and a total of 446 personnel;
  - The 193st Special Designation Battalion with 140 men and 8 BTR-80s; and
  - The 451st Operational Designation Regiment with two battalions, an air defense battalion, 28 BTR-80s, and a total of 600 men.

From Vladikavkaz—Grouping 2

under the command of the Deputy Commander of the Airborne Forces
General-Lieutenant Chindarov

- **Battalions—11** (motor rifle—1, tank—1, airborne—4, internal forces—5)
- **SP bn, Arty bn—2, indep. anti-air bn—2, SAM btry**
- **Personnel—3,915**
- **Helicopters—14**
- **Tanks—34, BTRs—67, BMPs—98**
- **Weaponry—62**

The Vladikavkaz grouping included the following forces:

- From the Northern Caucasus Military District:
  - Forces of the 19th Motor Rifle Division, including a motor rifle battalion, a tank battalion, an artillery battalion, an air defense battalion, 28 tanks, 34 BMPs, 10 units of weaponry, 6 ZSU-23/4 Shilka systems, 4 S10 SAMs, and a total of 723 personnel;
  - The 3/481 SAM Regiment of the 19th Motor Rifle Division including 4 Osa (SA8 Gecko) SAMs and 42 personnel; and
  - The 1/933 Independent SAM Regiment of the 42nd Army Corps including 4 Kub missiles and 36 personnel.
- From the Paratrooper Forces:
  - The 76th Paratrooper Division including 3 paratrooper battalions, an SP battalion, 41 BMPs, 12 units of weaponry, 30 units of air defense weaponry, and 1,125 personnel; and

- The parachute battalion of the 21st Independent Paratrooper Brigade, including 3 airborne companies, a howitzer battery, an anti-tank battery, 10 units of weaponry including grenade launchers, 6 units of air defense weaponry, and a total of 350 personnel.

- From the Internal Affairs Ministry Forces:
  - The 46th Operational Designation Regiment including 2 battalions, a SAM battery, a Spetsnaz company, 29 BTRs, 18 BMPs, and a total of 639 personnel;
  - The 7th Spetsnaz Detachment including 6 BTRs and 168 personnel; and
  - The 47th Operational Designation Regiment including two battalions, two independent anti-air battalions, 30 BTR-80s, 21 BMPs, and a total of 650 personnel.

From Kizliar—Grouping 3

under the command of the 8th Guards Army Corps Commander General-Lieutenant Rokh lin, L.Ya.

- **Battalions—8** (motor rifle—2, internal forces—6)
- **SAM bn, anti-air arty bn, indep. anti-air bn—2, howitzer btry—2, missile btry**
- **Helicopters—15**
- **Personnel—4,053**
- **Tanks—7, BTRs—162**
- **Weaponry and mortars—28**

Finally, the Kizliar grouping:

- From the Northern Caucasus Military District:
  - Forces of the 20th Motor Rifle Division including an independent reconnaissance battalion, a missile battery, 2 howitzer batteries, a SAM battery, a missile battery, a SAM battalion, 83 BTRs, 29 units of grenade launchers and other weaponry, 15 units of air defense weaponry, and 1,712 personnel.
- From the Internal Affairs Ministry Forces:
  - The 57th, 63rd, and 49th Operational Designation Regiments including 6 battalions, 3 Spetsnaz companies, and 2 SAM battalions.

Overall Complement of Forces

- **Battalions—34 (motor rifle—5, tank—2, airborne—7, internal forces of the Internal Affairs Ministry—20)**
- **Artillery bn—4, SP bn—2, SAM bn—3, anti-tank btry—2, howitzer btry—4, missile btry**
- **Helicopters—90, including 47 attack**
- **Personnel—nearly 24,000, including**
  - Armed forces of the Russian Federation—19,000
  - Internal forces of the Internal Affairs Ministry—4,700
- **Tanks—80, BTRs and BMPs—208**
- **Weaponry and mortars—182**

The three groups together comprised the total force shown above.

The inner and outer blockade called for by the plan was to be effected as follows: The outer ring was the responsibility of Internal Affairs Ministry forces. They were to block the main rail and automobile routes so as to prevent Chechens from neighboring areas, such as Dagestan, from entering Chechnya. Note that while Dagestan remains Russian territory, it has a significant ethnic Chechen population. The units were deployed as follows:

From the East:

- The 81st Operational Designation Regiment at Petropavlovsk on the left bank of the Argun River
- The 57th Operational Designation Regiment at Nizhni Gerzel', Gerzel'-Kutan

From the West:

- The 59th Operational Designation Regiment at Zebir-Urt and Dolinskiy
- The 22nd OBRON at Novi Redant

From the South:

- The 47th Operational Designation Regiment from Nazran to Assinovska
- The 46th Operational Designation Regiment at Bamut, Novi Shatoi, and the eastern edge of Samashka

From the North:

- The 451st Operational Designation Regiment from Brastoye and Verhniy Naur
- The 21st OBRON at the eastern edge of Galyugaevskiy and Lenposelok
- The 51st Operational Designation Regiment at Podgornoye and Ken'-Urt

The inner ring around Grozny was to be effected by the 19th Motor Rifle Division, the 76th and 106th Paratroop Divisions, the 131st Independent Motor Rifle Brigade, and the 21st and 56th Independent Paratrooper Brigades.

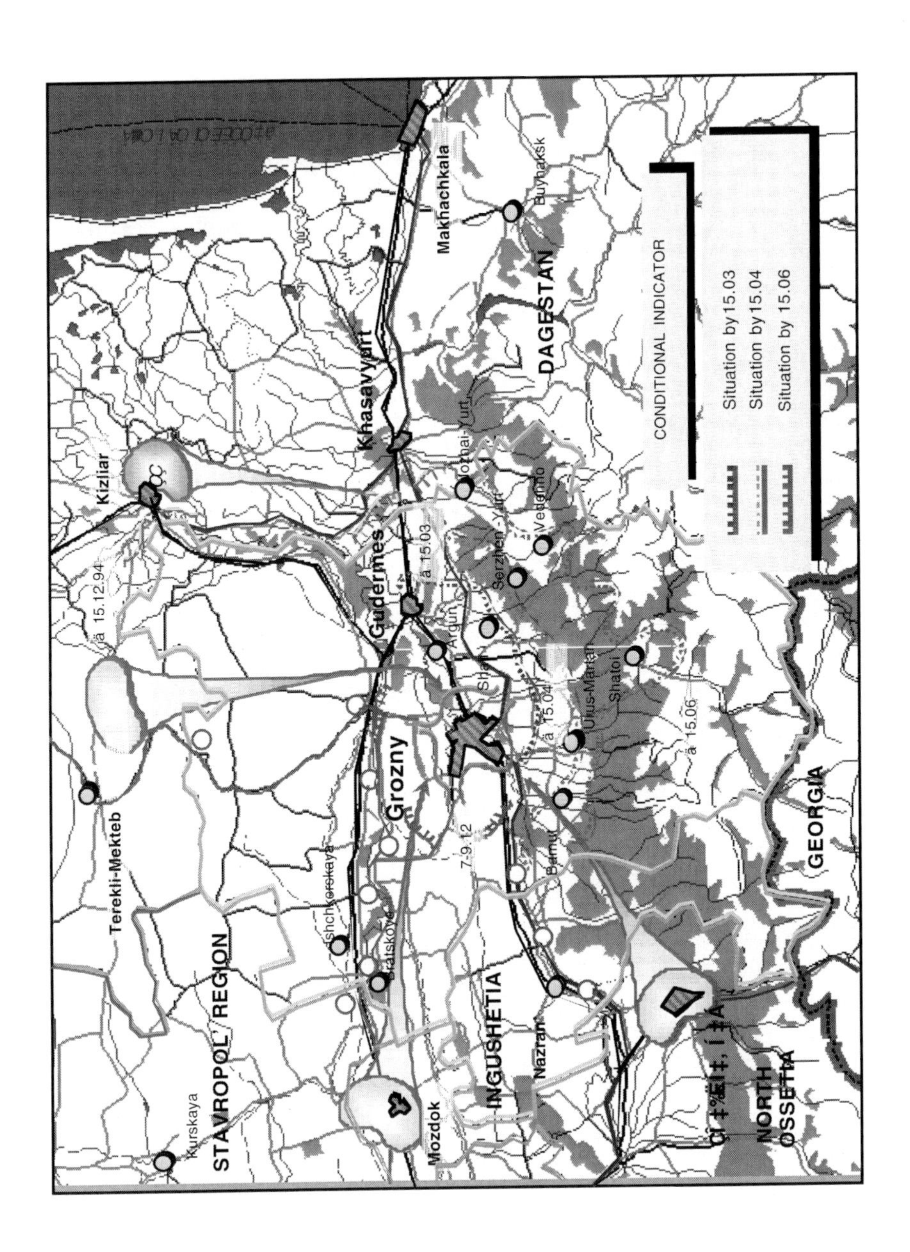

Kizliar
Khasavyurt
Makhachkala
Buynaksk
DAGESTAN
CONDITIONAL INDICATOR
Situation by15.03
Situation by15.04
Situation by 15.06
Gudermes
á 15.03
á 15.12.94
Grozny
Argun
á 15.04
Urus-Martan
Shatoi
á 15.06
Bamut
7.9.12
Terekli-Mekteb
Bratskoye
STAVROPOL REGION
Kurskaya
Mozdok
INGUSHETIA
Nazran
NORTH OSSETIA
GEORGIA

- First (Mozdok) axis: The 131st MRB was to move on the city from a position 4 km southeast of Terskaya. The paratrooper regiment and battalion, respectively, of the 106th PD and the 56th PB were to approach from the Mozdok airfield.
- Second (Vladikavkaz) axis: The 693rd MRB of the 19th MD was to begin from a position 1 km east of the Chermensk circle, the paratrooper regiment of the 76th paratrooper division from 3 km northeast of Beslan, and the paratrooper battalion of the 21st Independent Paratrooper Brigade from 2 km south of Kantishevo.
- Third (Kizliar) axis: The 20th Motor Rifle Division moved toward Grozny from the northern edge of Averyanovka.

Force movements were thus to follow six separate paths

- Mozdok, Bratskoye, Znamenskoye, Nadterechnoye, Kem'-Urt, Pervomayskoye
- Mozdok, Predgornoye, Novi Redant, Garagorsk, Kerla-Urt, Pervomayskoye
- Chermen, Verkhniye Achaluki, Karabulak, Sernovodsk, Alkhan-Kala, Alkhan-Urt
- Chermen, Gamurgiyevo, Northern edge of Yandirka, Novi Shatoi, Alkhan-Urt
- Kizliar, Hamamat-Urt, Nizhni Gerzel', Gerzel'-Aul, Novogroznensiy
- Terekeli-Mekteb, Baklan, Lugovoye, Chervlennaya uzlovaya, vinogradnoye, Petropavlovskaya

The first four of these routes covered a distance of 80–90 kilometers each, the last two 110–190 kilometers.

The operational plan was approved by the National Security Council of the Russian Federation on November 29, 1994.

Because the operation was to be carried out on the territory of the Northern Caucasus Military District (NCMD) and would utilize a large number of its forces and units, responsibility for the details of operational planning, preparation of forces, and command of troops was assigned to the NCMD leadership. Officially, this was formulated in Ministry of Defense Directive No. 3 12/1/00148 Sh of November 30, 1994. The NCMD was to report on plans for the operation on December 5. In order to assist with planning, organization, coordination of forces, troop preparation, and command and control, special Ministry of Defense operational groups were created.

On December 5, 1994, the Minister of Defense was briefed on the NCMD plans for the operation and approved them. He also directed that NCMD commander General-Colonel A. N. Mityukhin was to serve as the commander for the Joint Grouping of Forces. He was to have full command and decision authority over preparation for and implementation of the operation, including personnel issues, and he was to have this command and decision authority over all forces, including those of non-MOD ministries involved in the operation.

The Minister of Defense also directed that night operations were to be avoided in the interest of force protection and the security of the local population. He furthermore outlined the following schedule:

- All operational planning was to be complete by 1400 on December 6 as was force deployment to the region.
- Force groupings were to be formed by December 7 and communications training in all areas was to be carried out by 1600 on that date.
- Finally, force inspections were to be held on December 8.

This agenda was formalized by MOD Directive No. 3 12/1/002 Shr.

Per instructions from the President, it was also determined that the Minister of Defense would meet with representatives of the opposition forces and personally with D. Dudayev. These meetings were held on December 6 with unfavorable results.

The operation began on Sunday, December 11, 1994 at 0800 (rather than 0500 as the Ministry of Defense had decided the previous day). General-Colonel A. Mityukhin asked Minister of Defense Grachev at 2330 on December 10 to let the hour slip as the Vladikavkaz grouping was not yet ready to move. The delay resulted in a loss of any element of surprise. By 0900 the Vladikavkaz grouping had found its way blocked by unexpected and unarmed opposition from the local residents of Ingushetia in the region of Nazran and Verkhni Achaluk. (Note that the Ingushetians have close ethnic ties to the Chechens.) Furthermore, the troops found their way blocked by an automobile market where by 0800 thousands of people had assembled to buy and sell vehicles near their starting point in Nazran. As a result, it took them nearly two hours to even reach the border with Chechnya.

The fact was that Russian force and unit commanders proved unprepared to take decisive action. Only forces moving from Mozdok and Kizliar were able to keep to the schedule initially, and only one of the six groups moving along their separate routes was actually able to maintain that schedule to reach its position on the Grozny perimeter at the planned time. These were the forces moving from Mozdok in the north. The other groups only reached their positions by the 20th and 21st of December. As a result the blockade of Grozny was never complete. The south of the city, in particular, remained open.

Forces and Weaponry of D. Dudaev

- **Personnel—up to 10,000**
- **Tanks—25**
- **BTRs, BMPs—35**
- **Weaponry—up to 80 (D-30 122mm howitzers)**

The decision to enter the city of Grozny was taken at a National Security Council meeting on December 26, 1994. The rationale was that the majority of rebel forces and a significant portion of their armament and materiel were concentrated in Grozny. Weapons supplies were also located in the city. The illegal armed bands, despite repeated calls on them to end their resistance, continued intensive armed attacks, effecting a partial redeployment of their forces to previously prepared bases in the south of the republic.

But Dudaev's allies paid particular attention to preparing for the defense of Grozny, where, not counting those forces withdrawing to the south or the local population, some 9,000–10,000 personnel were stationed. Chechen equipment included about 25 tanks, 35 BMPs and BTRs, and about 80 ground artillery systems (mostly the D-30 122mm howitzer).

To effect the defense of the city, the rebel command hastily prepared three concentric defensive rings.

- An inner ring, centered around the Presidential Palace, had a radius of 0.5–1.5 km.
- A middle ring, at a distance of about 1 km from the inner ring in the northwest and up to 5 km in the southwest and southeast.
- An outer ring along the perimeter of the city and stretching to Dolinsk.

The rebels created knots of resistance around the Presidential Palace for the inner ring of defense, making use of buildings in the vicinity. Lower and upper floors of those buildings were adapted for use as firing positions for personnel armed with rifles and anti-tank weapons. Positions for direct artillery and tank fire were prepared along Ordzhonikidze and Victory Roads and Pervomaysk Street.

The foundation of the middle ring defense was strong points at the beginning of Staropromislovsk Way, knots of resistance at the bridges across the Sunzhe River and in the area around Minutka Square as well as on Saihanov Street, and preparations to set fires and/or explosions at the Lenin and Sheripov oil processing factories and other oil-related commercial enterprises as well as at the chemical factory.

The outer ring of defense consisted of strongpoints on the Grozny, Mozdok, Deliysky, Katayama, and Tashkala expressways as well as in the Grozny suburbs of Neftyianka, Khankala, and Staraya Sunzhe in the east, and Chernorech'ye in the south.

In summary, the scope of preparations to defend the city of Grozny made it clear that D. Dudaev and his allies were unlikely to voluntarily give up their weapons and dissolve the illegal armed bands. Thus, in order to fulfill the demands of the presidential decrees, government statements, and decisions of the Security Council, the Commander of the Joint Grouping of Federal Forces had only one option: to capture Grozny and begin the process of disarming Dudaev's illegal armed bands and associated foreign mercenaries in the Chechen capital itself.

The plan to capture the city began, of course, with its blockade. However, as already noted, delays of most of the forces in approaching the city meant that the blockade was never accomplished. Only the northern portion was in any way covered. The south of the city remained, for all practical purposes, completely open. If the failure to move at 0500 on December 11 was the first mistake made by the Russian forces, the failure to initially wait until the blockade was complete was the second mistake. But the decision was eventually made to attack the city regardless.

The operational plan called for the separation of Grozny into areas or zones, with the railroad tracks and the Sunzhe River serving as boundaries in the east-west and north-south directions respectively. Storm detachments were to attack from several directions at once: the north, the west, and the east as indicated. Upon entering the city they were to coordinate with special forces of the Ministry of Internal Affairs and the FSB to capture the Presidential Palace, other selected government buildings, television and radio stations, the train station, and additional important sites in the city's center and then blockade the central part of the city and the Katoyama region.

The attack from the north was to be carried out by two storm detachments from the 81st Motor Rifle Regiment of the Northern Force Grouping and a storm detachment of the 255 Motor Rifle Regiment of the 20th Motor Rifle Division of the northeast grouping, attacking along a zone bordered by Yuzhnaya, Mayakovskoyo, Krasnoznamennaya, and Mira Streets on the right and the Sunzhe River on the left. In this way they were to block off the northern part of the city and the Presidential Palace from the north.

Furthermore, the 131st brigade, attacking along Mayakovskogo Street and then along Staropromislovkogo Road, was to take up defensive positions along the Altaiskaya Street and on the border between the Staropromyslovsk and Leninsk regions of town.

From the west, two storm detachments of the 19th Motor Rifle Division (Western Force Grouping) were to attack along a zone bordered on the right by the railroad tracks and on the left by Popovicha Street. They were to capture the train station, then, moving north, to blockade the Presidential Palace from the south.

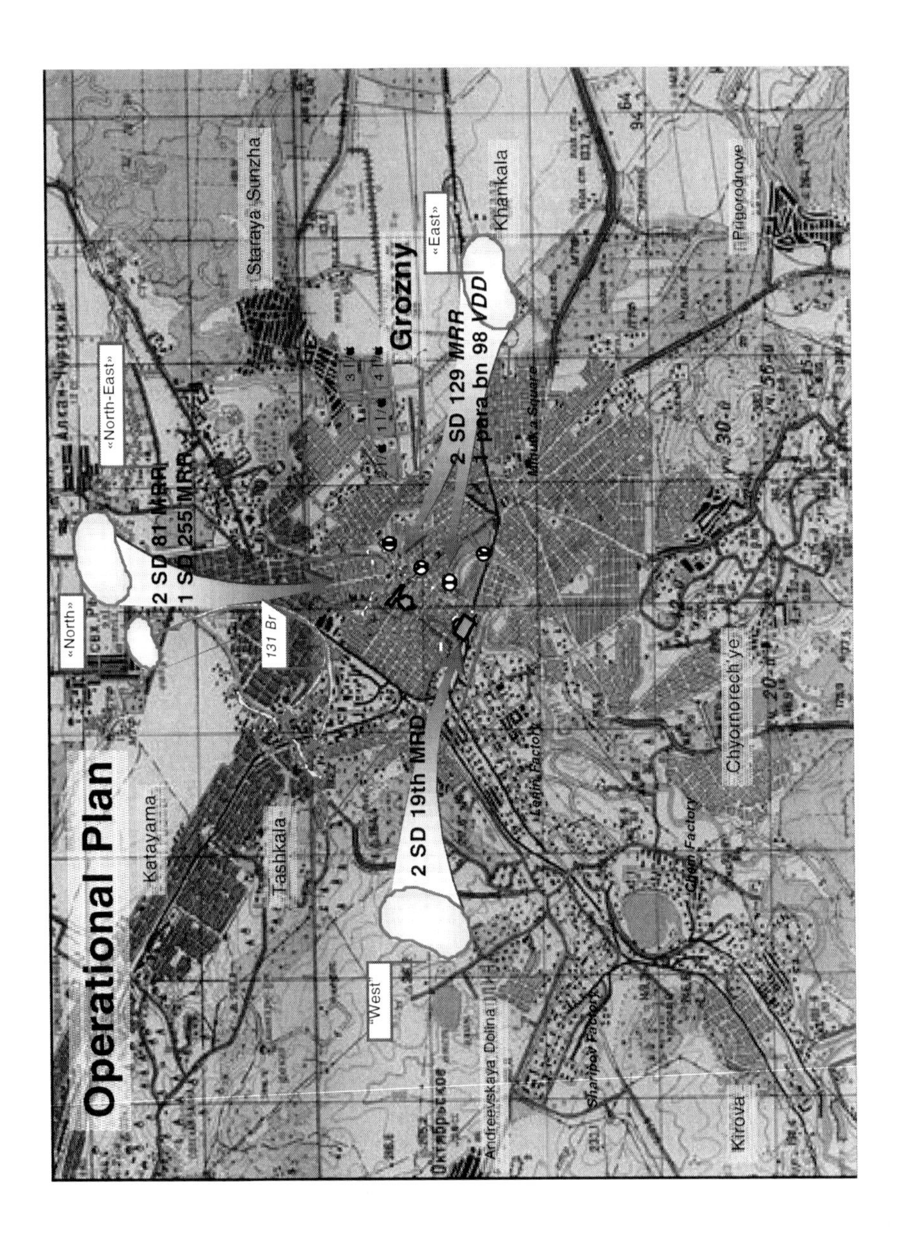
Operational Plan
Katayama
Tashkala
«North»
«North-East»
Staraya Sunzha
Grozny
«East»
Khankala
Prigorodnoye
2 SD 81 MRR
1 SD 255 MRR
131 Br
2 SD 129 MRR
1 para bn 98 VDD
2 SD 19th MRD
"West"
Andreevskaya Dolina
Chyornorech'ye
Kirova
Lenin Factory
Chem Factory
Sharipov Factory
Minutka Square

The combination of these attacks and blocking of major streets was to create a corridor in a zone bordered on the right by Bohdan Khmelnitsky, Pervomaysk, and Ordzhonokitze Streets and on the left by the Sunzhe River (and further west by Lenin Park).

In order to prevent combat in the western part of the city where the chemical and petroleum processing complexes were located, and to prevent enemy movement into the rear of our forces, detachments of the 76th and 106th Paratrooper Divisions were to block off the Zavod and the Katoyama regions.

From the east, two storm detachments of the 129th Motor Rifle Regiment and a paratrooper battalion of the 98th Paratrooper Division (all from the Eastern Grouping) were to move along the railroad tracks from Gudermes to Lenin Square and then to the Sunzhe River. There they were to capture the bridges across the river and link up with the Northern and Western Force Groupings to block off the central part of the city and the river from the east.

Finally, federal troops had an additional reserve composed of the 131st Motor Rifle Brigade and the 8th Motor Rifle Regiment. These were assigned to the Northern Group, which was ultimately to have the most success in accomplishing its tasks. They had a unique mission: to isolate the rebel formations in the northwest from the city proper.

The role of the Internal Affairs Ministry Forces, in turn, was to defend and guard communications and join the armed forces in disarming the illegal armed bands once the city was captured.

The expectation was that the federal forces, approaching a single point from three directions, would fully surround Dudaev's forces located in the center of the city. Casualties among the Russian troops would be minimized, as would collateral damage to the city of Grozny. But the success of this operation depended on the element of surprise.

There was a contingency plan in the event that Dudaev's forces presented active resistance. Russian forces were to capture the city over the course of several days. This contingency was the reason that the storm detachments had been created. Those forces had even received additional training to prepare them for their mission. This training was not, however, sufficient to overcome the fact that these detachments were composed of subcomponents and even individuals from a range of different divisions and forces who had never before fought or trained together. This significantly hampered operational effectiveness.

The attack on the city began on December 31st. The Western Grouping met with resistance in the industrial region of the city and found itself engaged in close combat. The Eastern Grouping encountered resistance and decided to take a detour, maneuvering to the right. However, they encountered additional resistance and minefields along this new route. In the end they were forced to retreat. The Northern Grouping was able to approach the Presidential Palace but encountered fierce resistance near that building. Thus, at 1500 on December 31st, all three groupings had failed to attain their goals. Recall, however, the role assigned to the 131st Motor Rifle Brigade. They were supposed to move south along with the rest of the Northern Grouping, but then turn west to isolate rebel forces in the northwest. The brigade's commander met with no resistance during the move south but failed to make the turn, instead continuing south. He reported his situation and was told to continue on to the train station. He and his forces did so, again encountering no resistance. Upon arriving at the train station, however, the brigade failed to carry out the key tasks of securing the area, encircling the station, and posting guards at strategic elevated positions in nearby multistory buildings. Instead, the 131st MRB was extremely careless, leaving their BMPs and many of their weapons in the square in front of the station while most of the personnel congregated inside the building. As a result they were easy prey for the rebel forces that soon surrounded and attacked them. Most of their weapons and BMPs were lost in the fight and the majority of their vehicles were destroyed. I personally saw the remains of a tank close to the train station. It had been demolished, taking some 20 direct RPG hits.

With it increasingly clear that Russian plans for the Grozny operation were not going to be successful, Prime Minister Chernomyrdin held a meeting on January 9 with the Minister of Defense in attendance. I stated openly at this time that until we successfully sealed the city from the south we would be unable to capture it. The fighting in January caused a lot of casualties on the Russian side. Because the south remained open we were unable to stem the tide of rebel reinforcements. For practical purposes we were engaged in a battle with the whole of Chechnya. Still, we made some progress. The Presidential Palace was finally captured on January 19. On January 26, I was appointed commander of the Joint Federal Forces Grouping. At this time I was able to order the blockade of all of Grozny,

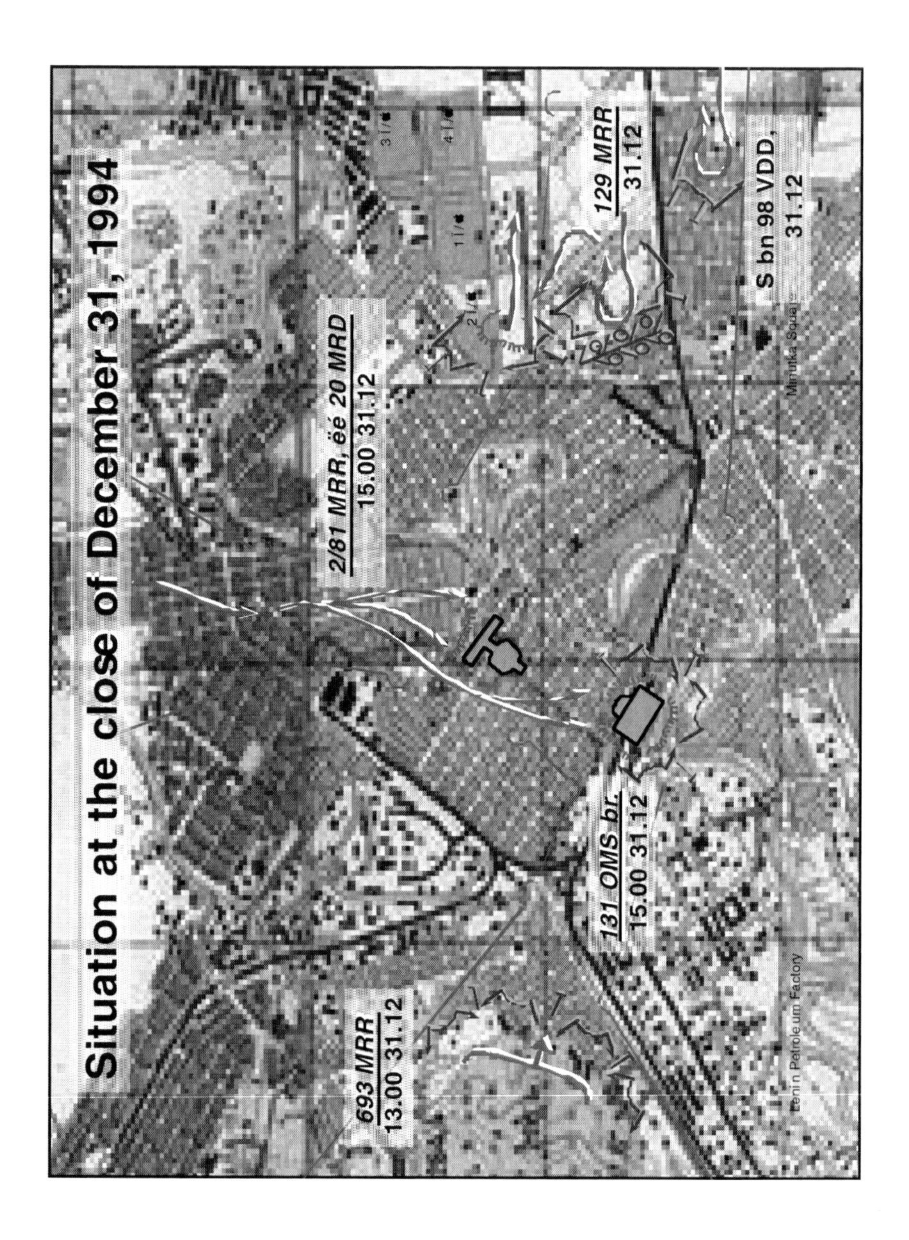
Situation at the close of December 31, 1994
2/81 MRR, ëé 20 MRD
15.00 31.12
129 MRR
31.12
S bn 98 VDD,
31.12
Minutka Square
131 OMS br.
15.00 31.12
693 MRR
13.00 31.12
Lenin Petroleum Factory

including effecting a seal in the south. Due to significant armored resistance we were unable to complete this mission until February 23, but in the end we were successful. Once the city was sealed, we were able to take control of it entirely within a week.

Grozny Operations 1999–2000: Lessons Learned?

This major error of the first war, the failure to implement a complete and effective blockade, was one of the lessons learned that was incorporated in operations in 1999. While MoD reconnaissance forces attempted to attack the city at the end of that year, they did not move until the city was isolated. Grozny was eventually captured in a campaign lasting from January 3 until February 7. This time Internal Affairs Ministry Forces took the lead with support from the Ministry of Defense. Tactics differed significantly from 1994. Four sniper companies were formed, two from the MoD and two from the Internal Affairs Ministry. These, supported by Spetsnaz forces, took control of key buildings within the city and operated at night with night vision goggles. They then supported the six storm detachments that entered the city by eliminating rebel forces when they attempted to approach Russian units.

The rebels, however, had also changed their tactics. They had prepared their underground communications networks and reinforced shelters, particularly those in basements. According to Russian officers who fought in the 1999–2000 Grozny campaign, these were set up to defend against air raids and aviation strikes; the basements were retrofitted and covered by concrete blocks which could be raised and lowered on jacks. In this way, the Chechens could lift the concrete blocks over their shelters when Russian forces approached and shoot at them, lowering the blocks once the troops had passed to protect those underneath from the subsequent aviation strikes called in by the Russians.

Federal Operations in the Plains

By the beginning of March 1995 the city was completely under Russian control. Having experienced defeat and heavy casualties in the battles for Grozny, the leadership of the illegal armed bands took time to regroup its forces. Dudaev's forces were generally well prepared for armed resistance in the east. Rebel bases in the south and southeast were in place and forces were supplied with weaponry and materiel. With the coming of warm weather, the guerrillas shifted their tactics to an emphasis on diversionary terrorist operations.

There was reason to think that the strategy of D. Dudaev and his allies had all along been simply to buy time until spring and summer when they could shift to partisan warfare. In the meantime, the combination of stoic resistance and diversionary terrorist acts throughout the entire territory of Chechnya kept the fight going.

By the middle of March we had captured Argun, Gudermes, and Shali and disarmed the illegal bands in those cities. By first surrounding the cities we were able to minimize Russian casualties in these operations. By focusing attention first on the plains prior to turning to the mountains, we were able to establish control over these areas, which comprised the majority of Chechnya's territory, by April 15. The rebel forces were largely disbanded and a large portion of their heavy weaponry (which they were unable to replace without foreign assistance) was destroyed. Their command and control system was compromised. Russian forces for the most part had complete control over the "Northern Path" rail line between Mozdok, Chervlenaya, Gudermes, and Grozny in particular.

Engineer forces in Grozny demined and checked for explosives throughout the city, paying particular attention to facilities necessary to everyday life such as the water supply system and key administrative buildings. Over 70,000 explosives were removed and destroyed, not including ammunition and guns. In sum, 770 hectares of territory were cleared.

In this way, favorable circumstances were created for a return to everyday life in the Chechen Republic and in Grozny.

However, federal forces still had significant work left to do in disarming illegal bands in the Samashka, Achkhoi-Martan, and Bamut areas. When completed, this operation would provide federal control over the "Southern Path" of the rail line to Grozny.

Federal Operations in the Mountains

In order to disarm the illegal bands in the mountain areas in the south of the Chechen Republic, border troops moved in the western (including Achkhoi-Martan, and Shatoi regions as well as parts of Ingushetia), central (Shatoi region), and eastern (Vedensk and Nozhait-Yurtovsk regions) directions to seal the borders.[2] In addition, a force grouping of armed forces and Internal Affairs Ministry forces was created to clear out the sealed-off area. The grouping commenced its work in the summer.

In the beginning of the summer, the Dudaev regime had sufficient forces and means to maintain partisan warfare, even with losses. They were able to do this due to the support that continued to exist from the local population as well as that received from representatives of the new administration of the Chechen Republic and other northern Caucasus republics.

The plan for the summer operation in the mountains called for it to be completed in 20 days.

In the course of operations in May and June of 1995, the illegal armed bands suffered significant casualties and were brought to the edge of complete destruction. Remaining guerrilla forces were forced to retreat to eastern regions of the republic near the towns of Dargo and Venoi and to southern territories near the Dagestan border, as well as east towards the town of Bamut. Guerrillas, understanding the futility of further resistance, began to leave the armed bands.

However, following Sh. Basaev's attack on Budennovsk, morale in the illegal armed formations rose significantly. Guerrillas began to return to their units, new formations were created, and new volunteers joined up. Basaev's success at Budennovsk and the support for it amongst many of the field commanders convinced the Dudaev leadership of the need to shift to a new tactic of armed resistance in which diversionary-terrorist acts would be carried out not just on the territory of the Chechen Republic but in central areas of Russia as

[2]Russian forces hoped to make it difficult for the rebels to maneuver in the mountains by destroying and mining roadways. They also used motorized rifle, tank, and paratrooper forces to destroy remaining weaponry and personnel of the illegal armed bands.

well. Furthermore, the Dudaev leadership did not reject the possibility of continuing to resist federal forces in defensive battles in zones of contact.

The Russian political command, and particularly Prime Minister Chornomyrdin, had made a grave mistake in its response to the Budennovsk situation. First, they entered into negotiations with the terrorists. Then they suspended military actions, which prevented the military from completing its job.

All in all, however, the thinking behind operations to disarm the bands in the mountain regions of the Chechen Republic was consistent with the goals and conditions of the situation. The major failing was that the planners did not anticipate the need for a reliable reserve force, an error that had significant operational impact.

Conclusions

One of the key lessons learned from our experience in planning the operations to disarm illegal armed bands in the Chechen Republic was our recognition of just how crucial it is to take into account all of the factors that can impact force operations: the need for a joint command, unity of command for all forces, the fact that we were conducting operations on Russian territory, the existence of constrained timelines for preparation and planning, and particularly the need to form and deploy forces.

Operational planning was conducted with the historical experience of the Second World War in mind. The overall plan was developed by the General Staff; detailed planning was the responsibility of the leadership of the Joint Federal Force Grouping that included officers of the Ministry of Defense, General Staff, and representatives of other ministries and organizations whose forces and components were participating in the operation.

One of our most significant problems turned out to be the organization of a single unified command for the forces. Incompatibility between different forces' technical equipment made it difficult to create a single system and impeded operations. Military operations in Chechnya also underscored the continuing discrepancy between the potential of space systems and the limited resources in this sphere due to our current force infrastructure.

Harsh time constraints dictated by the operational plans made it impossible for traditional reconnaissance methods to provide timely information to commands and forces. In fact, the army and federal command had almost no reconnaissance whatsoever prior to the beginning of operations. Security and Internal Affairs Ministry forces similarly could not carry out reconnaissance as they had no access to the area before operations began. Such problems can only be effectively solved with the use of space-based reconnaissance, communications, and command and control systems. Joint Grouping forces must plan in advance to ensure that they are supplied with compact systems for reception of information from space-based reconnaissance assets, especially those supporting high-precision weaponry.

Problems of poor unit and personnel readiness further complicated preparation for operations. This demonstrated the need to improve not only coordination between forces in the theater but also the training of personnel. This occupied a significant amount of our time (ten or more days). Our experience highlighted the need to train and prepare personnel at their home base locations to make maximum use of the educational base in place there. Of course, the fact that operational units were created ad hoc by combining soldiers and components from a range of commands and organizations had a negative impact as well.

The rebel forces enjoyed a number of advantages over our troops. These included:

- Familiarity with the territory on which operations were conducted
- Support of the local population
- The ideological factor (based in Islamic nationalism)
- The use of professional mercenary troops
- Significant anti-tank assets
- Active use of snipers
- Effective use of defense against maneuver

The rebels also suffered from a number of weaknesses, however, including:

- A drop in morale after their defeats in the first stage of the conflict
- Dissent and disagreement amongst the military and administrative leadership
- Increasing dissatisfaction with Dudaev's regime
- Lack of educated personnel
- Dependence on outside sources of financing for ammunition, supplies, and other materials.

THE CHECHEN
OPERATION FROM
THE VIEWPOINT OF THE
MILITARY COMMAND

BACK-UP SLIDES

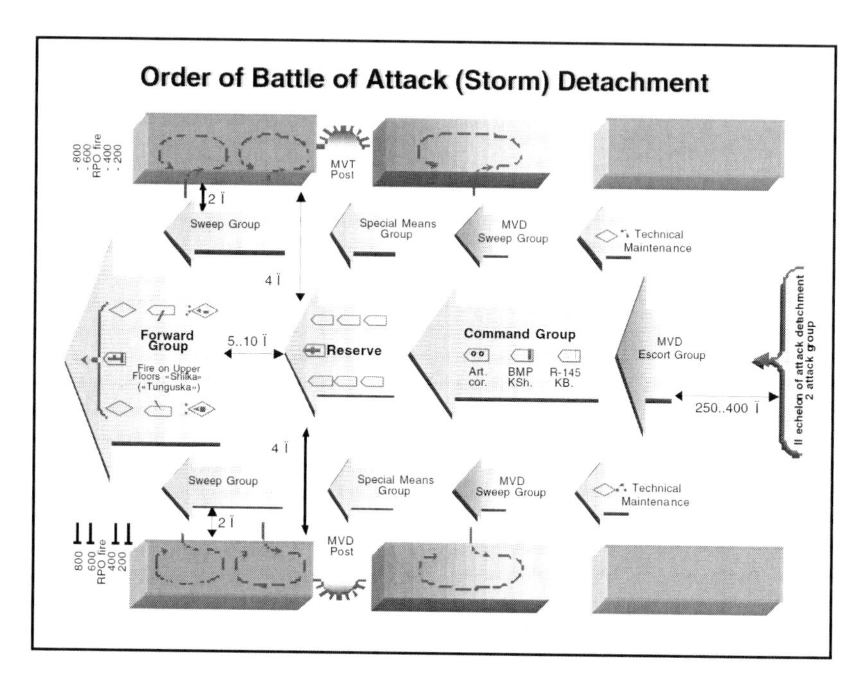

Order of Battle of Attack (Storm) Detachment
MVT Post
Sweep Group
Special Means Group
MVD Sweep Group
Technical Maintenance
Forward Group
Fire on Upper Floors «Shilka» («Tunguska»)
5..10 Ï
Reserve
Command Group
Art. cor.
BMP KSh.
R-145 KB.
MVD Escort Group
250..400 Ï
II echelon of attack detachment 2 attack group
4 Ï
2 Ï
MVD Post
800
600
RPO fire
400
200

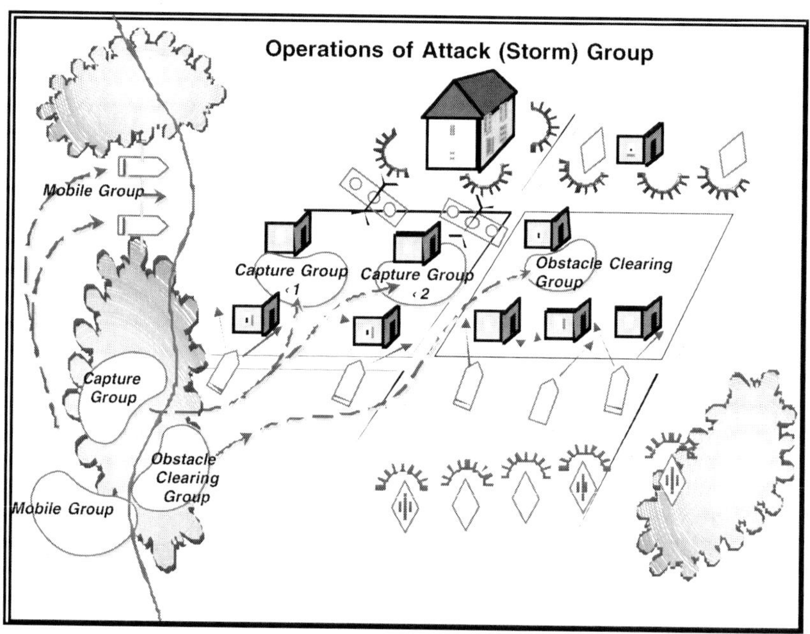

Operations of Attack (Storm) Group
Mobile Group
Capture Group ‹1
Capture Group ‹2
Obstacle Clearing Group
Capture Group
Obstacle Clearing Group
Mobile Group

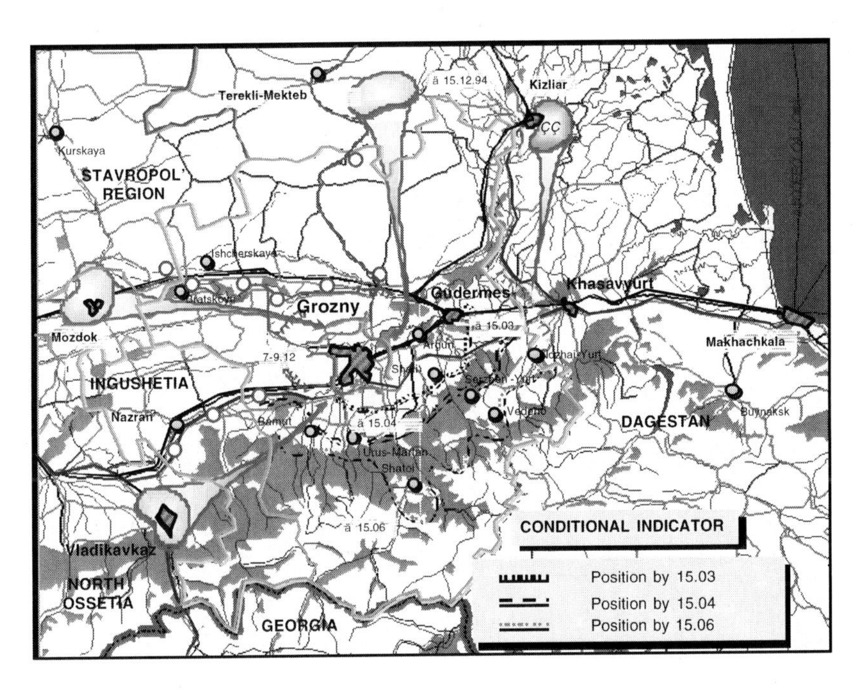
Terekli-Mekteb
ä 15.12.94
Kizliar
Kurskaya
STAVROPOL' REGION
Ishcherskaya
Gudermes
Khasavyurt
Grozny
Mozdok
ä 15.03
Argun
Makhachkala
7-9.12
Shali
INGUSHETIA
Serzhen-Yurt
Nazran
Vedeno
DAGESTAN
Buynaksk
Bamut
ä 15.04
Urus-Martan
Shatoi
ä 15.06
CONDITIONAL INDICATOR
Vladikavkaz
Position by 15.03
NORTH OSSETIA
Position by 15.04
GEORGIA
Position by 15.06

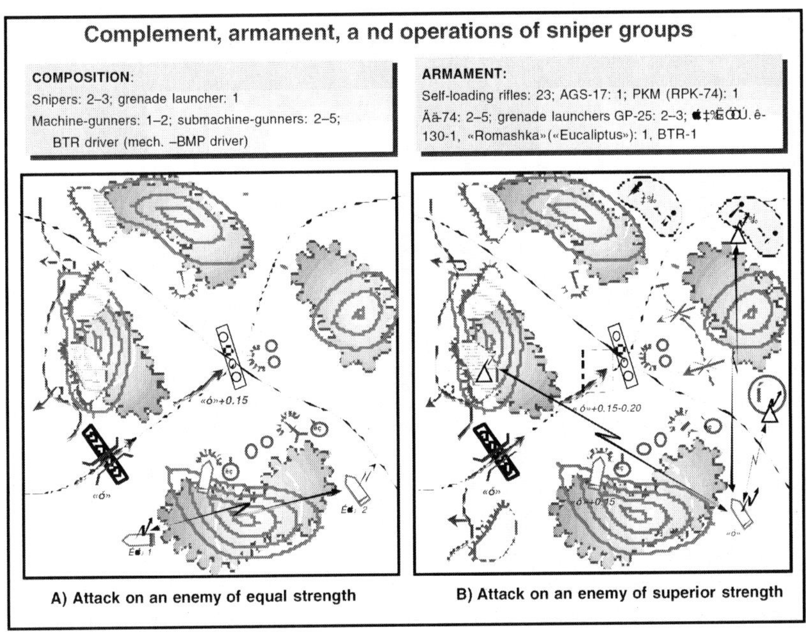
Complement, armament, a nd operations of sniper groups
COMPOSITION:
Snipers: 2–3; grenade launcher: 1
Machine-gunners: 1–2; submachine-gunners: 2–5;
BTR driver (mech. –BMP driver)
ARMAMENT:
Self-loading rifles: 23; AGS-17: 1; PKM (RPK-74): 1
Ää-74: 2–5; grenade launchers GP-25: 2–3;
130-1, «Romashka»(«Eucaliptus»): 1, BTR-1
«ó»+0.15
«ó»
«ó»+0.15-0.20
A) Attack on an enemy of equal strength
B) Attack on an enemy of superior strength

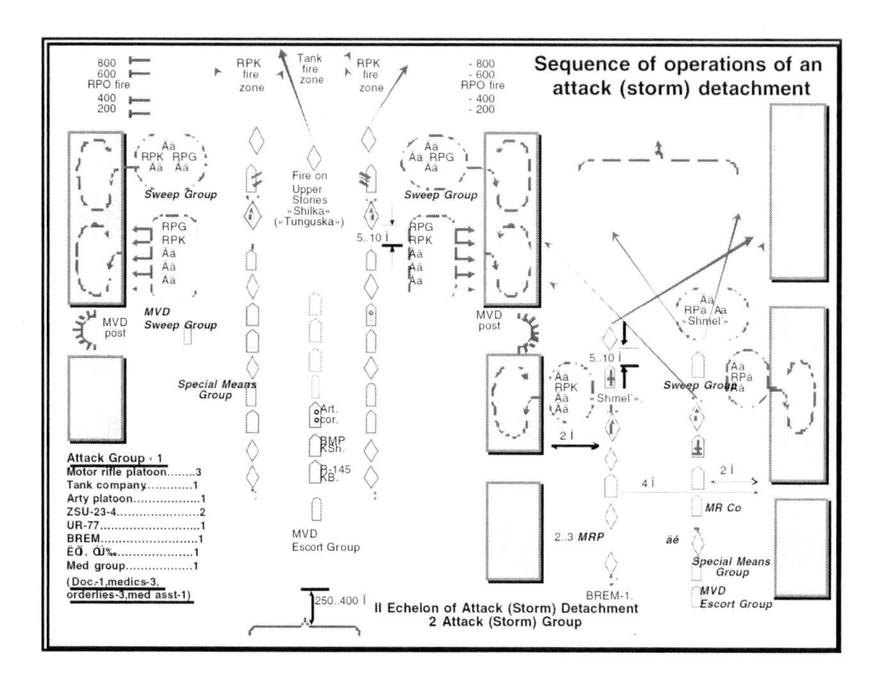

Sequence of operations of an attack (storm) detachment
800
600
RPO fire
400
200
RPK fire zone
Tank fire zone
RPK fire zone
- 800
- 600
RPO fire
- 400
- 200
Sweep Group
Fire on Upper Stories «Shilka» («Tunguska»)
Sweep Group
RPG
RPK
5..10
MVD post
MVD Sweep Group
MVD post
Special Means Group
Shmel'
Sweep Group
Art. cor.
BMP KSh.
R-145 KB.
2
4
2
MR Co
Attack Group ‹ 1
Motor rifle platoon........3
Tank company............1
Arty platoon.................1
ZSU-23-4......................2
UR-77..........................1
BREM...........................1
Med group..................1
(Doc-1,medics-3, orderlies-3,med asst-1)
MVD Escort Group
2..3 MRP
Special Means Group
MVD Escort Group
BREM-1.
250..400
II Echelon of Attack (Storm) Detachment
2 Attack (Storm) Group

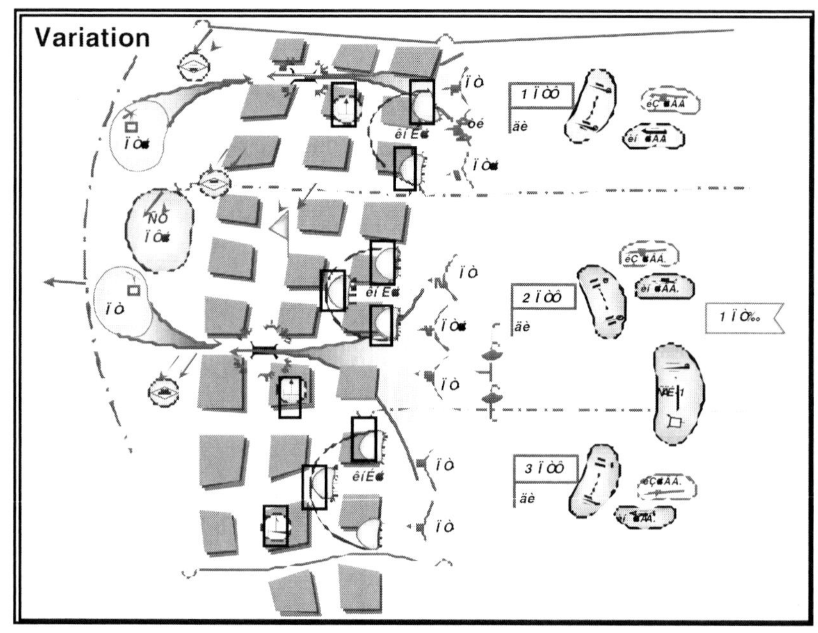

Variation

Dependence of the Level of Casualties on the Degree of Destruction by Fire of an Object During Its Capture

Attacking forces and weaponry	Damage to target, %	Reduction in casualties, %	Saved forces and weaponry
Motor Rifle Battalion [Attack (Storm) Detachment]	10	15	Up to one Motor Rifle Platoon
	20	23	Up to one Motor Rifle Company
	30	60	Up to 1.5 Motor Rifle Company
Motor Rifle Companies (3); Tank Company (1); Artillery Battalion (1); Mortar Battery (1); Sapper Platoon (1); SAM Battery; Light Infantry Flamethrower Company (1)	40	80	Up to 2 Motor Rifle Companies
	50	88	Object capture can Become a task for an Attack (storm) group
	60	92	

Appendix C

THE TWO SIDES OF GROZNY

Arthur L. Speyer, III
Marine Corps Intelligence Activity

CHECHNYA:
Urban Warfare Lessons Learned

Marine Corps Intelligence Activity—Arthur Speyer

The strategies and tactics employed by the Chechen resistance in the battle of Grozny offer outstanding lessons for future urban operations. Grozny, the capital of the breakaway Russian republic of Chechnya, is the site of the largest urban warfare operation since the end of World War II. The Chechen resistance that continues to fight a prolonged conflict against Russian forces provides a model of the 21st-century urban insurgency. Chechen tactics are being studied by insurgent groups worldwide and may one day be used against U.S. forces in the streets of Kosovo, Bosnia, Indonesia, or Liberia.

Why Chechnya Matters

"The future of war is not the son of Desert Storm, but the stepchild of Chechnya."
–General Krulak USMC (ret.)

- Largest urban battle since World War II
- Russia employed all aspects of conventional military power within an ubran environment

The Chechens keenly demonstrated how a small, decentralized, and lightly-armed insurgency can defend against a larger, conventionally organized military in an urban environment.

The Chechen Resistance

- Textbook example of the modern urban guerilla
- Conducted asymmetric warfare
- Nationalistic/religious political
- Ties to international organizations

The Chechens knew they could not defeat the Russians in a direct conflict. To counter Russian strength, the Chechens attacked Russian weaknesses. They moved throughout the city and denied the Russians a true front line. The Chechens attacked again in the Russian rear and denied the Russians a decisive battle. The Chechens also used contacts abroad, mainly in the Middle East and Turkey, to acquire equipment and seasoned fighters.

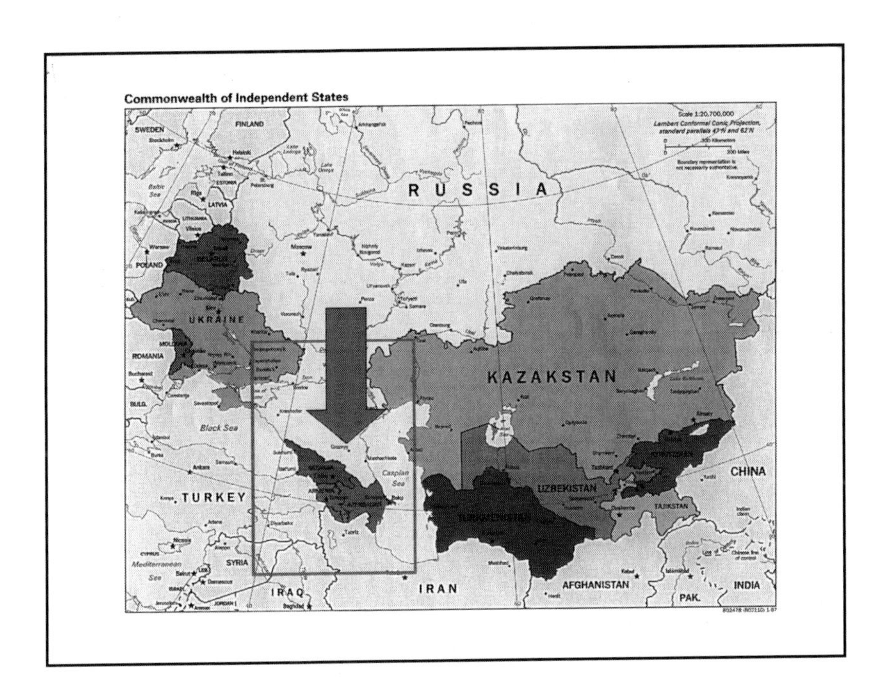

The Chechens are an Islamic, clan-based ethnic group that inhabits the mountainous Caucasus region of southern Russia, one of the most ethnically diverse areas of the world. The group is fighting for independence from Russia for a mix of political, cultural, religious, and economic reasons. The Russians and the Chechens fought two major battles for control of Grozny; one during the winter of 1994–95 and the second during the winter of 1999–2000.

Background to the Conflict

- Chechnya is part of the Russian Federation
- Long history of resisting Russian control
- Began latest drive for independence in 1990

The Chechen people have a long history of resisting Russian control. Following the collapse of the Soviet Union, they began in earnest to seek full independence. In 1994, Chechnya became a civil war battleground between pro-independence and pro-Russian factions.

In December 1994, Russia sent approximately 40,000 troops into Chechnya to restore Russian primacy over the breakaway republic.

Grozny—Pre-War

50,000 people

68 square miles

High rises
Industrial areas
Suburbs

Modern by Soviet standards

The Chechens

- Clan-based mountain people
- Long history of fighting Russian aggression
- Islamic
- Trader/dealer culture

Chechen Strategy

- Inflict Russian casualties
- Extend the conflict
- Attack in rear areas
- Bring the fight to advantageous geography

The Chechens had no illusions about fighting the Russians. They knew they could not conduct conventional combat operations against them and win. However, they believed they could inflict serious damage if they could draw the Russians into the urban environment. As the battles for Grozny progressed, the Chechens began to realize the distinct advantages the urban environment gave to their decentralized operations.

While the Russians concentrated on securing territory, the Chechens aimed to inflict Russian casualties and extend the conflict. Chechen leadership sought to cause one hundred Russian casualties daily; the Chechens believed if they could continue the war and inflict high numbers of Russian casualties, Russia would eventually pull out.

Russian Strategy

- Seize and hold key terrain
- Destroy Chechen resistance
- End war quickly

The Russian strategy was simple and direct: gain control over the territory of Chechnya and destroy Chechen resistance.

Order of Battle

Russians

45,000 men—1995

95,000 men—2000

Chechens

15,000 men

During the first war in 1995, the Russians deployed over 45,000 troops from across the wide range of Russian security services to the Chechen theater. In 2000, that number doubled to approximately 95,000 troops. Chechen numbers are hard to judge due to the large number of part-time fighters. The Chechens likely had 15,000 men under arms though they rarely had more than 2,000–3,000 fighters in the city of Grozny at any one time.

The Chechen Defense of Grozny

- Chechens planned city defense for two years
- City manager and city engineer involved

- Intimate knowledge of streets, buildings, and strategic intersections

The Chechens knew that the Russians would send armor units directly into Grozny to destroy the Chechen resistance. The Chechens used the urban geography to plan their defense. City officials who controlled Grozny's roads, telephones, and power advised the Chechen military leadership. The Chechens drew on their intimate knowledge of local streets, buildings, and key intersections to defeat their enemy.

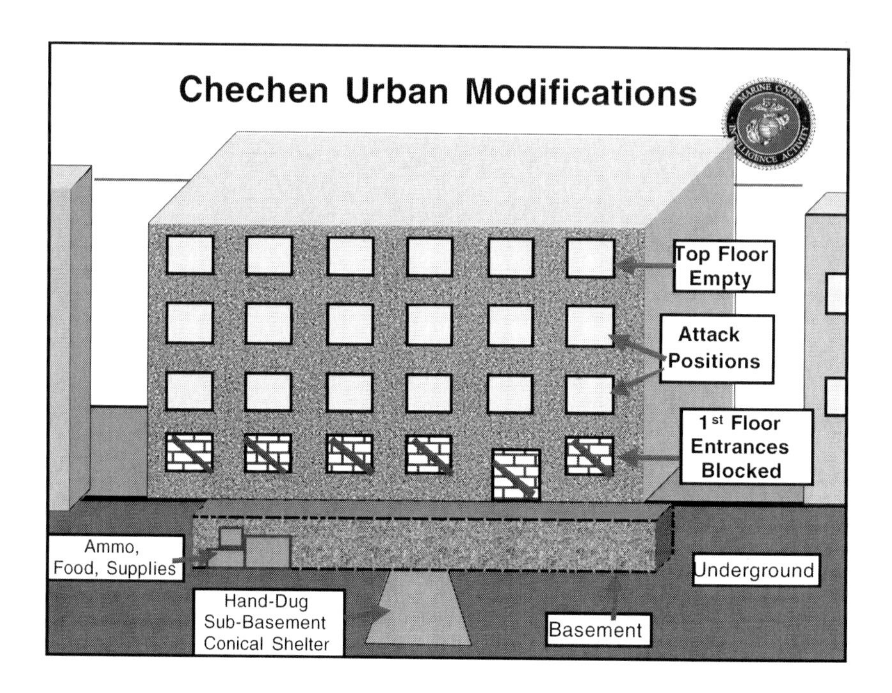

The Chechens would use urban structures as defensive positions. As the conflict progressed, they became expert at further modifying urban structures to give themselves extra protection against Russian forces. The Chechens would board up the first floor windows and doors to prevent their use by Russian ground troops. The top floor would be left empty for fear of Russian artillery or air attack. The Chechens would use basements and dig sub-basements to store supplies and survive massive Russian artillery strikes.

Initial Assault

- December 31, 1994
- City approached from east, west, and north
- Light resistance for Northern Group
- Northern Group convoys to center of city at dusk
- Long armored columns, no supporting infantry

After reaching the Chechen capital of Grozny, approximately 6,000 Russian soldiers mounted a mechanized attack into the urban area. The attack was launched simultaneously from three directions and featured tanks supported by infantry riding in BMP infantry fighting vehicles (IFVs). Instead of the anticipated "cake walk," Russian forces encountered heavy resistance from Chechen forces armed with large quantities of antitank weapons. The Russian attack was repulsed with shockingly high Russian casualties. It took another two months of heavy fighting and adapting Russian tactics to finally capture Grozny.

The Chechen Counter-Attack

- Attacked at sundown
- Chechens drew Russian forces into city center
- Chechen scouts monitored progress through suburbs
- Russian forces panicked

Russians lost 102 of 120 APC/IFV and 20 of 26 tanks

The Russians used a combination of vehicles during the first attack to include BMPs, BTRs, and MTLBs. The Chechens monitored Russian movement through the city using small, hand-held, off-the-shelf Motorola radios. They intentionally drew the Russians into urban canyons where they could ambush them and reduce Russian combat advantages.

Russian Combined Arms

STORM GROUP

- Motorized Infantry Company
- Tank Platoon
- Artillery Platoon
- Mortar Platoon
- AGS-17 Platoon
- Engineer Platoon
- Chemical Troops

Russian Infantry/Conscripts

- Basic fire and maneuver skills lacking
- NCO and junior officer leadership weak
- Many hid, ran, panicked, or deserted

- Young, unhealthy, untrained
- Over 2/3 had less than 6 months military experience
- Low morale

Infantry Fighting Vehicles

- High attrition rates
- Many lost at point-blank ranges to RPGs and heavy machine guns
- Increased ammunition needs due to nature of urban warfare

Air Defense Vehicles

- High-elevation weapons were excellent for urban combat
- 4,000–5,000 rounds per minute
- Excellent anti-sniper weapon

2S6

- Light armor proved vulnerable, became a prime target for Chechens

Russian Artillery

- Massed artillery used to devastate large areas of Grozny
- Used to compensate for poor infantry performance
- Russian artillery units outnumbered maneuver units around Grozny

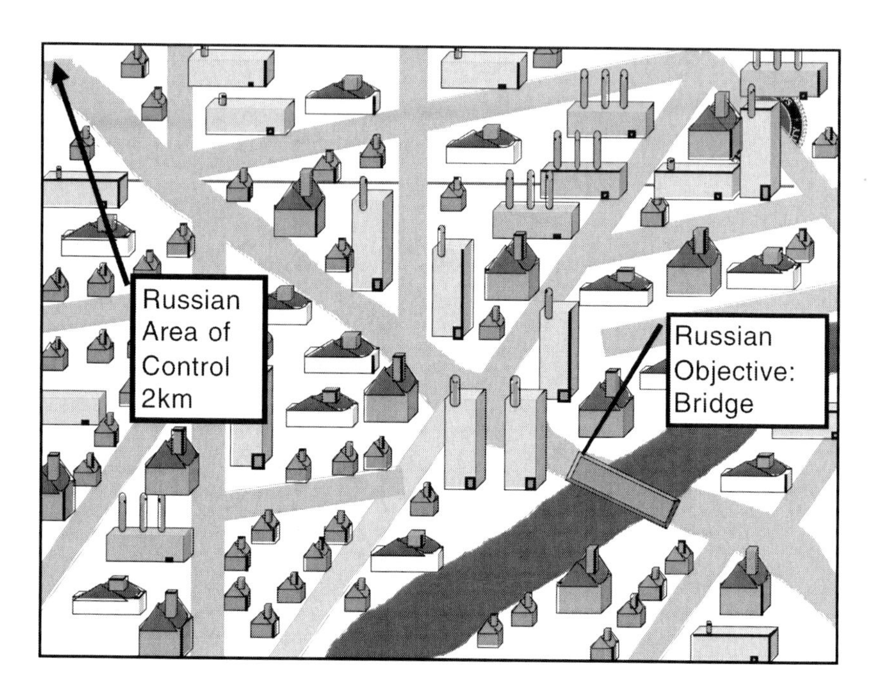
Russian
Area of
Control
2km
Russian
Objective:
Bridge

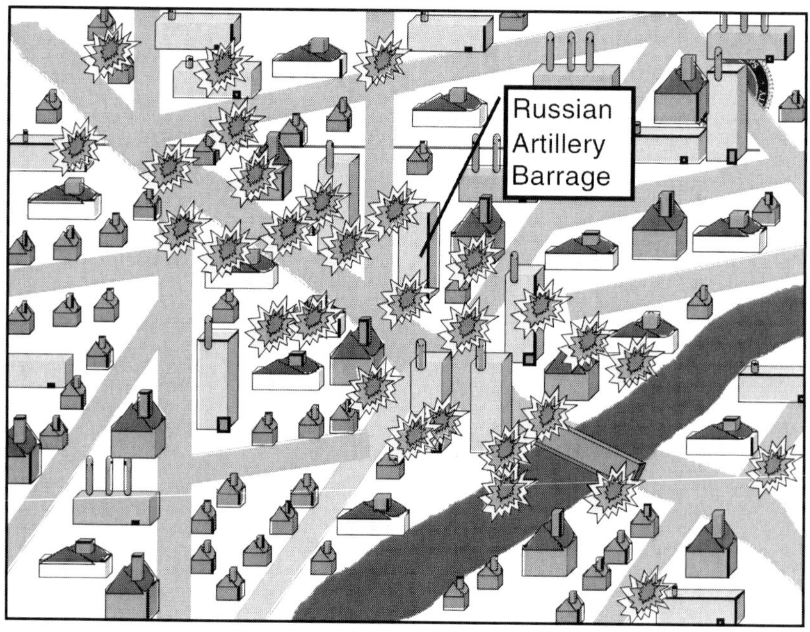
Russian
Artillery
Barrage

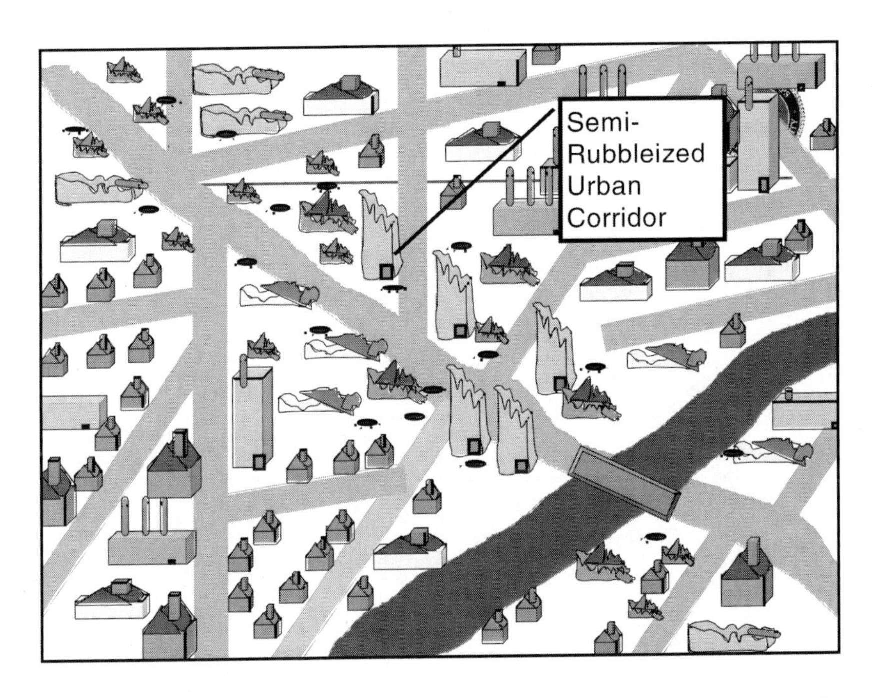
Semi-
Rubbleized
Urban
Corridor

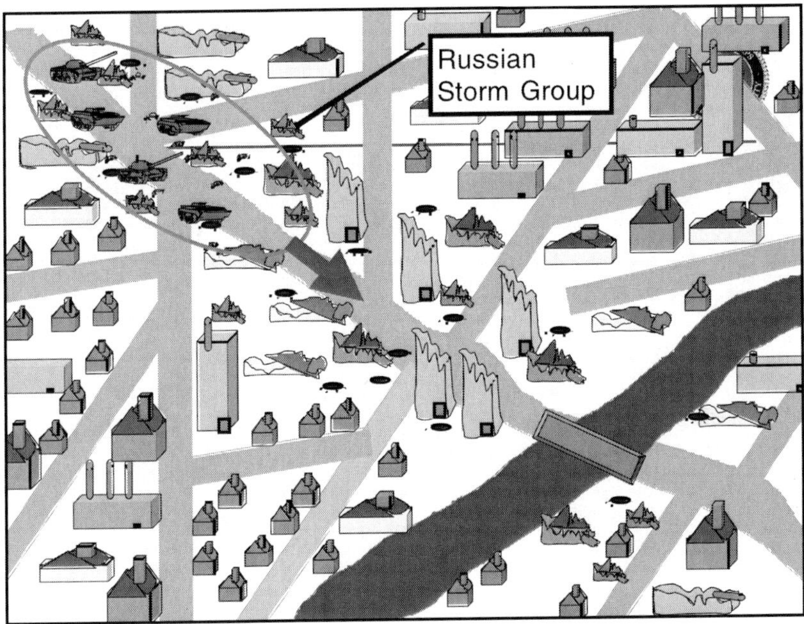
Russian
Storm Group

Intelligence

- Limited accurate maps
- Underestimated threat
- Cultural arrogance
- Did not conduct proper reconnaissance of city
- Value of imagery limited

Throughout the conflict, the Chechens had a well-developed human intelligence (HUMINT) network in the city. Chechen fighters rarely wore military uniforms and could easily blend into the city population. The locals were an excellent information source for the Chechens; they would routinely report Russian movements using small hand-held radios and couriers. Young women were particularly useful intelligence agents as they could easily move throughout the city.

Rotary Wing

- Employed weapons at stand-off range on city ring
- Would rarely venture into urban canyon for fear of attack
- Mainly used for logistic support

Airpower

- Hampered by poor weather
- Little direct support of ground troops
- Used free-fire zones
- Employed laser-guided bombs against high value targets

Why Chechnya 1999?

- Internal Security and the "Near Abroad"
- The military wanted to avenge loss in 1996
- Re-establish Russia as a legitimate superpower
- Stabilize/control oil region
- Fight "terrorism"
- Election year politics

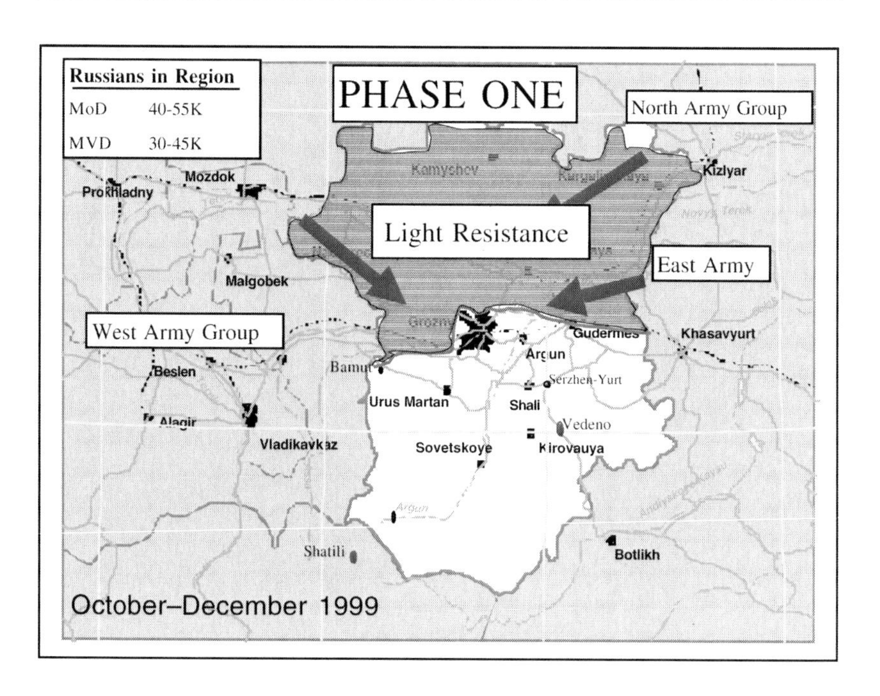

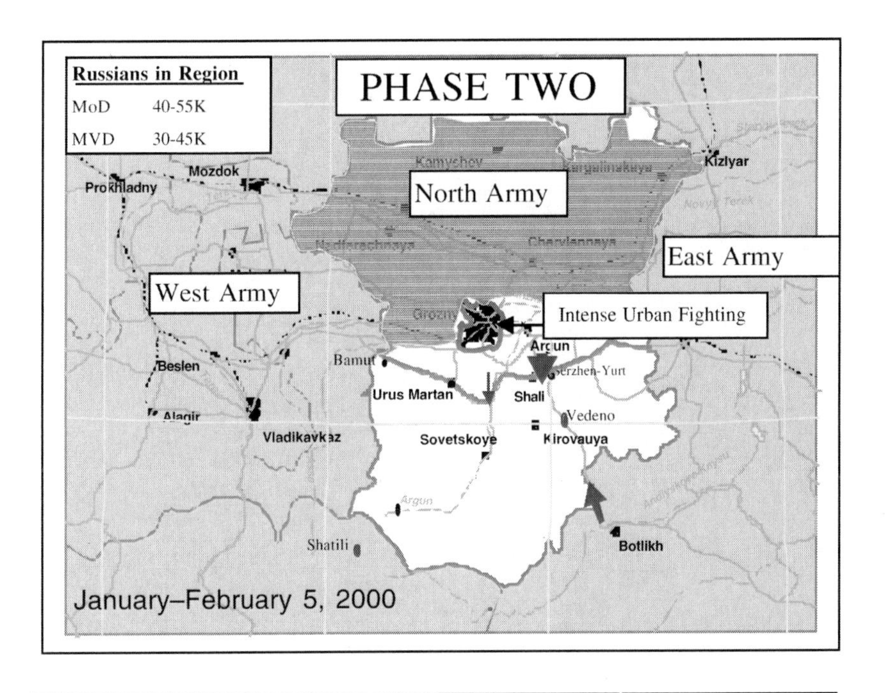

TOS-1 in Chechnya

- Mounted on T-72 chassis
- 220mm rocket tubes
- Unguided
- 4 warheads: napalm, FAE, thermobaric, thermite

Russian More-Than-Lethal

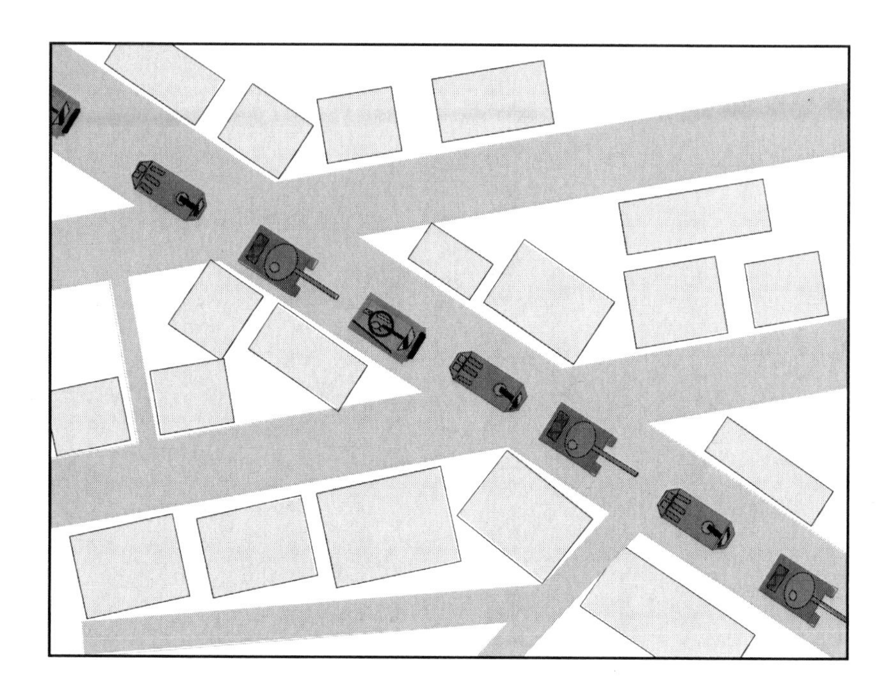

These graphics depict the Chechen counterattack on New Year's Eve 1994. The Russians were moving in a single column through the city with no supporting infantry. The Chechens attacked the front and rear vehicles. Once they were disabled, the other vehicles had little room to maneuver. The Chechens then moved down the line destroying the remaining Russian vehicles. The Russian forces panicked and could not launch a counterattack.

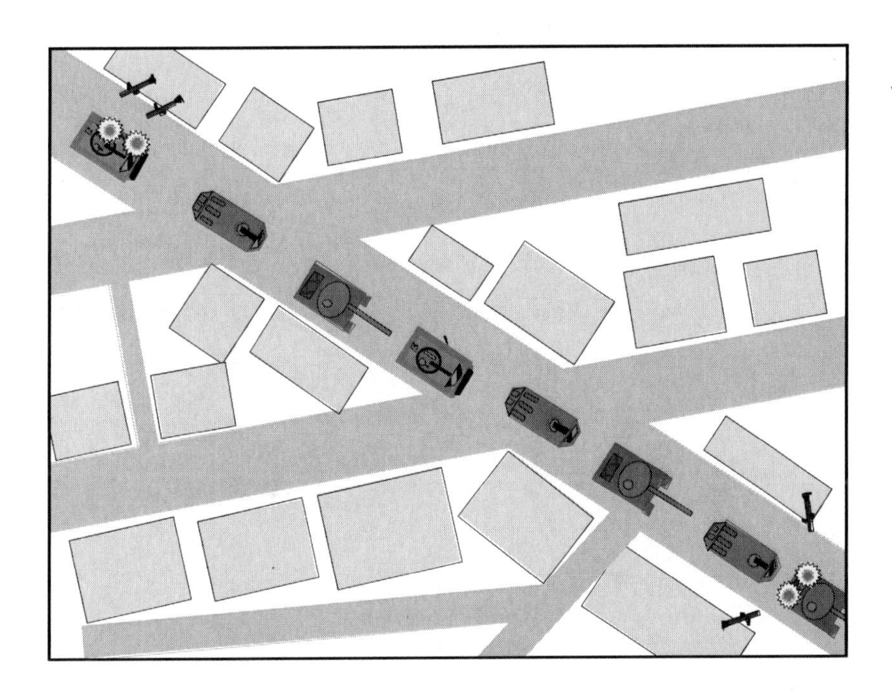

Chechen Urban Tactics

- Fought as light 25-man highly mobile teams
- Urban ambushes would rarely involve more than 75 fighters
- Would attack Russian units and break them into smaller pieces
- Owned the night

The Chechens made use of (and sometimes discarded) the often plentiful stock of captured Russian equipment. This included Russian Night Vision Devices (NVDs) for maneuver at night, maneuver often completed while under Russian bombardment. The Russians normally did not move at night or during periods of heavy fog while operating in the city. The Chechens used fog to mask their movement. They also keyed on the Russian use of smoke as an obscurant, taking it as an indicator of Russian movement. The Chechens would fire into the smoke with positive effect during Russian displacements.

Chechen Tactical Formations

8-Man Fighting Group

2 Heavy Machine Guns
2 RPGs
1 Scout/Sniper
1 Rifleman/Medic
1 Rifleman/Radioman
1 Rifleman/Ammo

25

8 8 8

Highly Mobile
No Body Armor

The Chechens centered their eight man subgroups (armor hunter-killer teams—squad equivalent) on the RPG ("Chechnya's national weapon"). Each subgroup contained three riflemen/automatic riflemen/ammunition bearers, two RPG gunners, one sniper, and two machine gunners. The sniper was also often employed as a spotter.

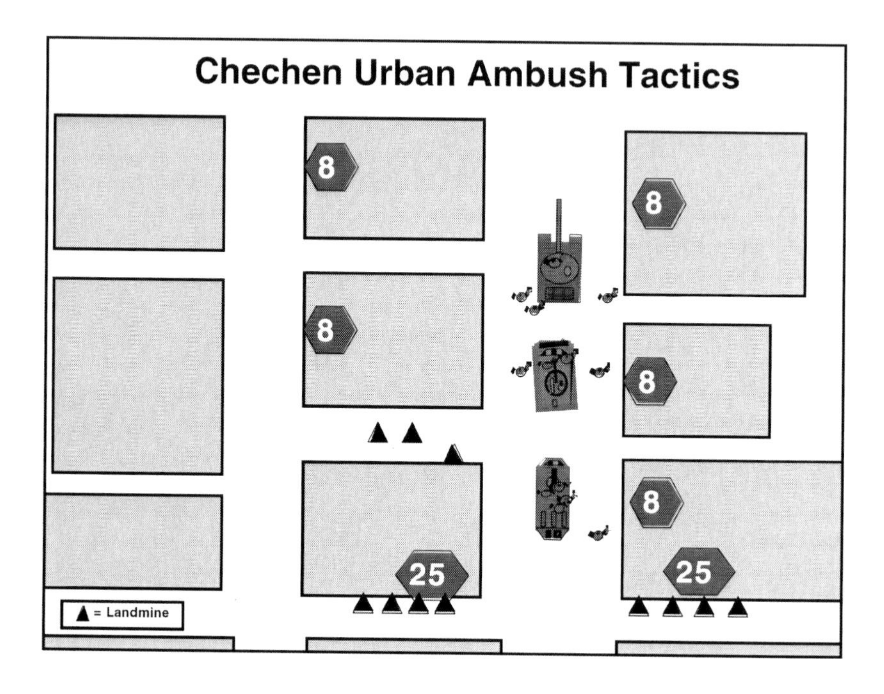

In the conduct of armor and personnel ambushes, the Chechens configured their forces into 75-man groups. These were further broken down into three 25-man groups (platoons). These platoons were further broken down into three equal-sized teams of six to eight fighters each (squads). Each squad had two RPG gunners and two PK (machinegun) gunners. The 75-man unit (company) had a mortar (82mm) crew in support with at least two tubes per crew. The Chechens did not move by flanking maneuvers against the Russians but instead incorporated chess-like maneuvers to hit them. They used buildings and other structures as navigation and signal points for maneuvering or initiating ambushes/assaults against the Russians.

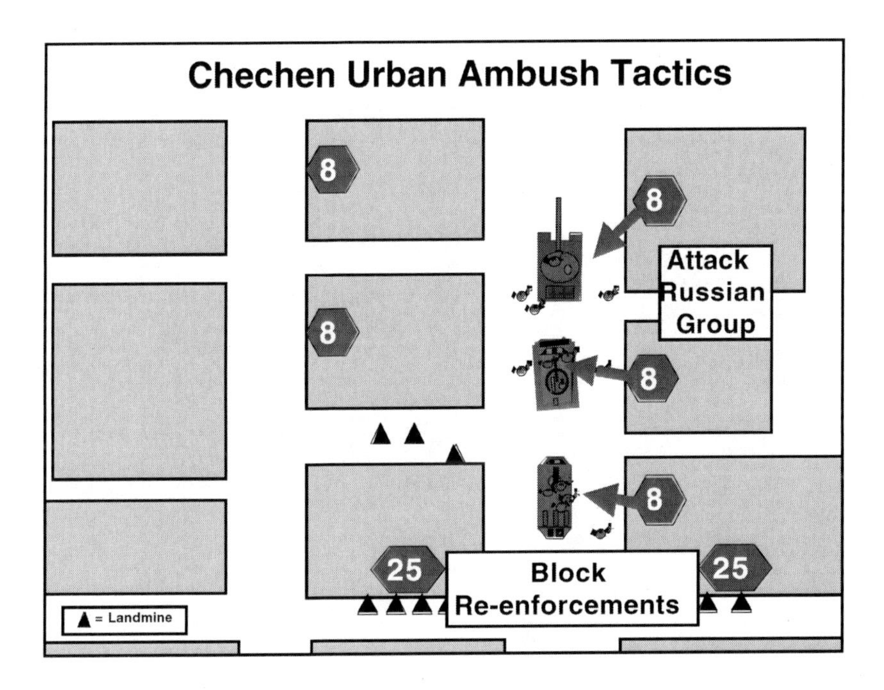

The Chechens only occupied the lower levels of multistory buildings to avoid casualties from rockets and air-delivered munitions coming through the upper levels. One 25-man platoon comprised the "killer team" and set up in three positions along the target avenue. They had responsibility for destroying whatever column entered their site. The other two 25-man platoons set up in the buildings at the assumed entry points to the ambush site. They had responsibility for sealing off the ambush entry escape or interdict efforts to reinforce the ambushed unit.

Chechen "Hugging"

- Stayed close to Russian infantry in urban areas, usually less than 50 meters
- Would rather take chances with infantry than suffer from Russian artillery or air

The Chechens utilized "hugging" techniques to reduce casualties from indirect fires. They would set up positions within 25 to a maximum of 100 meters of Russian positions in order to render Russian artillery and rocket support ineffective.

Chechen Infrastructure

- Relied heavily on stolen, captured, or purchased Russian supplies
- Every 8-man team had a Motorola hand-held radio
- Constructed extensive bunkers throughout the city

The primary communications device used by the Chechens was a small hand-held Motorola radio. It was used at all levels below "Headquarters" (national equivalent). At this higher level they had access to INMARSAT for communications with the outside world but kept these communications to a minimum because of the monetary cost involved. The Chechens had a ratio of about six combatants to each Motorola radio—but had they been able to afford more radios they would have issued every fighter one during the conduct of urban operations. The Chechens did not use any encryption or separate tactical nets. They maintained communications security by using their native language. Every Chechen could speak Russian but few Russians understood Chechen.

The National Weapon of Chechnya: The RPG

- Widely used in Grozny
- Two dedicated RPG gunners per 8-man team
- Disable first then destroy
- Engaged armor with multiple weapons
- Best RPG gunners 13–16 years old
- Effectiveness greatly increased in urban environment

The standard Russian Rocket-Propelled Grenade (RPG) warhead needed four rounds on target to penetrate a tank—the Chechens altered the RPG-7 round by removing the detonator cap and increasing the explosive components in such a way that they could penetrate a tank's (to include the T-72) armor and "blow the turret off" in one shot. The Chechens found the RPG to be an extremely effective weapon in urban warfare. Its simple use and wide availability earned it the nickname "The National Weapon of Chechnya." The Chechens employed the RPG against a wide range of targets, including bunkers, vehicles, personnel, and buildings. The RPG was successfully fired from multiple launchers against Russian armor. The Chechens targeted the thinly armored areas in the rear, top, and sides of Russian vehicles.

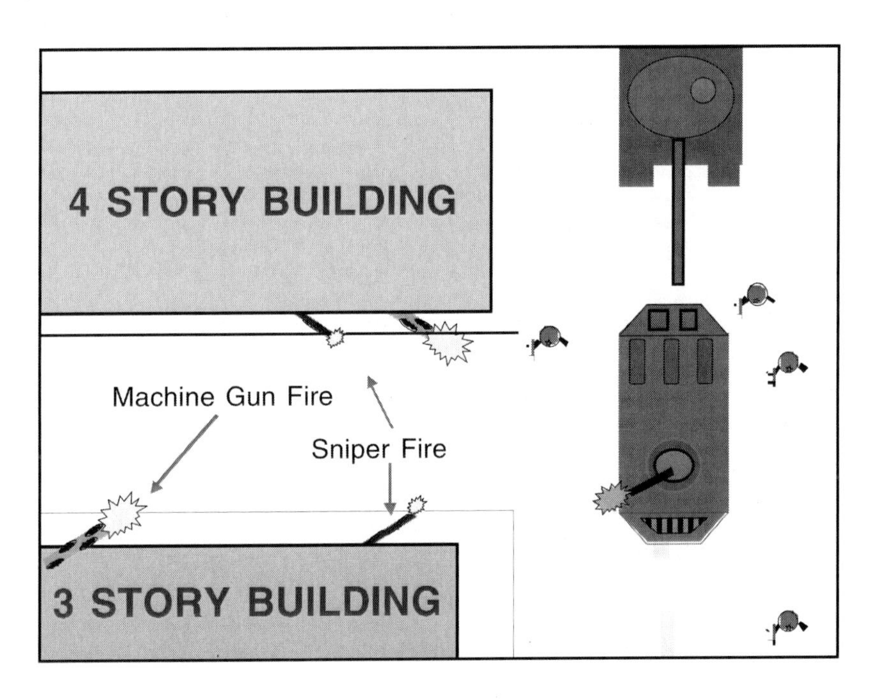
4 STORY BUILDING
Machine Gun Fire
Sniper Fire
3 STORY BUILDING

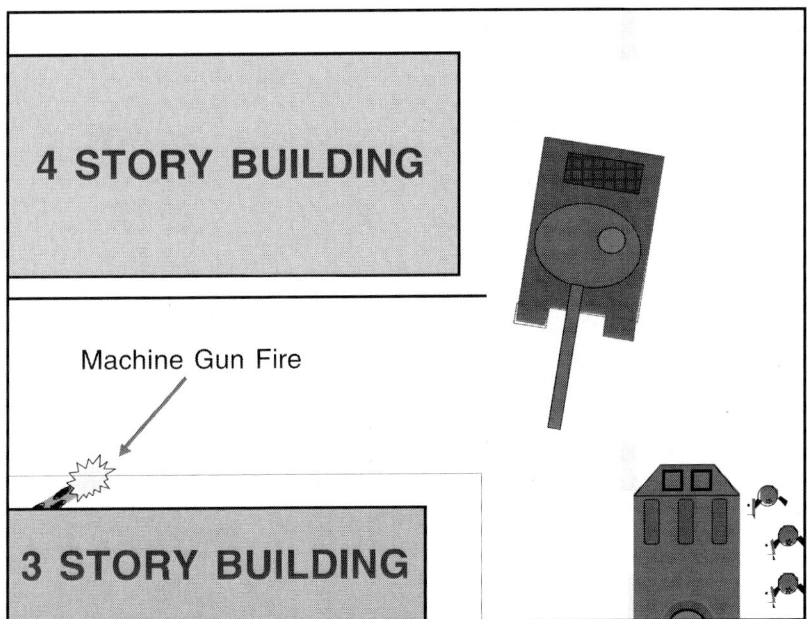
4 STORY BUILDING
Machine Gun Fire
3 STORY BUILDING

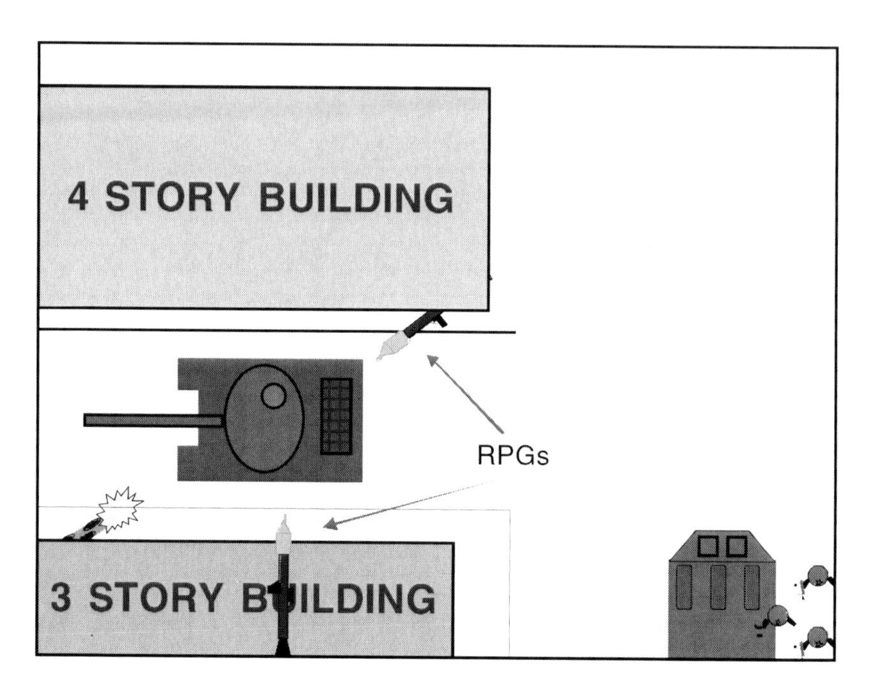
4 STORY BUILDING
RPGs
3 STORY BUILDING

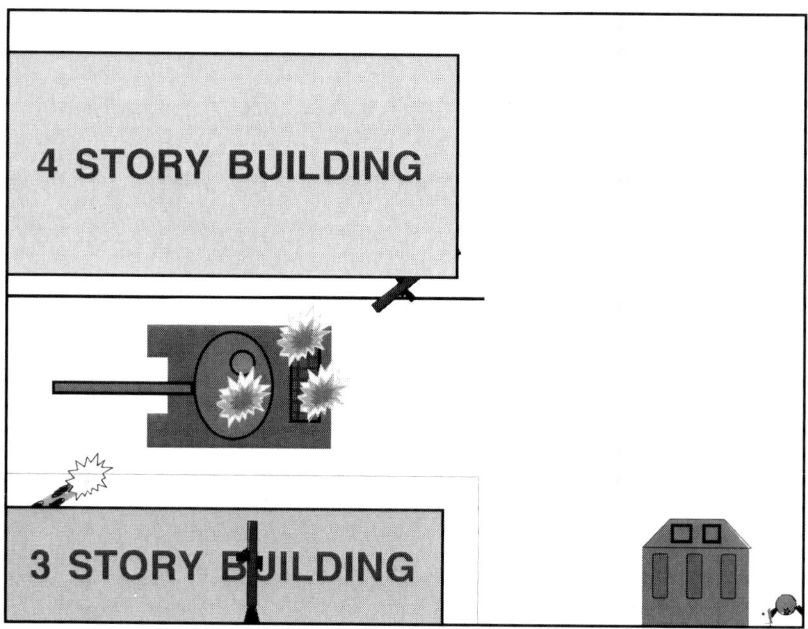
4 STORY BUILDING
3 STORY BUILDING

Urban Sniping

- Widely used by both sides in Grozny
- Second most deadly urban weapon for the Chechens
- Used as scouts as much as for sniping
- Russians employed in support of infantry; Chechens also depl oyed independently
- Best source of tactical intelligence in urban area

The most effective Chechen weapon system employed against "pure" Russian infantry was the SVD sniper rifle employed by a trained Chechen sniper. The SVD was not only effective as a casualty producer, but also as a psychological weapon that reduced morale among Russian ground troops. The Russians diverted significant combat power to search for Chechen snipers but were unsuccessful. A major reason for this was that the Chechens had prepared infantry positions to provide supporting/covering fire against Russian forces engaged in countersniper operations.

Chechen Weaknesses

- Internal divisions hindered effort
- Could not conduct an extensive engagement
- Many part-time fighters
- Limited supplies of ammunition
- Could not replace large battlefield losses

Despite tactical success in the streets of Grozny, the Chechens had several weaknesses. The Chechens' greatest weakness was their inability to conduct an extensive engagement. The small size of the Chechen units, coupled with their limited ammunition supplies, caused them to avoid large-scale battles. The Russians discovered that the Chechens could be defeated by forcing the rebels into drawn-out engagements, surrounding their positions, and using overwhelming fire support. The Chechens were unwilling to sustain the level of battlefield losses the Russians experienced and would retreat following a strong Russian counterattack.

The Chechens are a clan-based culture that only bands together to fight foreign invasion. Without a common enemy, the Chechens often turn against each other. The Chechen General Staff did not have complete control over all Chechen forces. Many were independent groups that decided themselves when, where, and how long they would remain in combat. The Russians never learned to successfully exploit these vulnerabilities and simply viewed all Chechens as terrorists.

Chechens with Stingers?

This is an example of possible Chechen deception or information warfare. It was found on a Chechen Web site and shows several Chechens in possession of Stinger anti-aircraft systems. The Chechens may or may not have working Stinger systems, but the possibility of their having them affected the way Russians employed their aircraft during the conflict.

Lessons from Chechnya

- The urban environment leveled the playing field between forces
- Detailed planning was essential
- Prior urban warfare training was critical to battlefield success
- Russians did not understand unique aspects of urban warfare (urban canyon, RPGs, etc.)
- Russians failed to understand cultural aspects

The first thing you must do—and it is priority number one—is study the people. You must know the psychological makeup of not only the combatants you might face but that of the local populace as well. Understand your enemy in detail—but not only from a military and political perspective—but also from a cultural viewpoint. If you underestimate the importance of this, you are on a road to decisive defeat. The Russians—given 400 years of conflict with the Chechens—have not learned this lesson. It is a matter of understanding your foe's mentality.

Lessons from Chechnya (Cont.)

- The key to victory in the streets of Grozny was the human will to fight
- Troops must be prepared to withstand long periods of intense combat in a chaotic environment with limited resupply or rest
- Technology and equipment were secondary

While it is unlikely the United States will conduct an operation similar to the Russian invasion of Chechnya, there are lessons of value that can be learned. Victory in the streets of Grozny was based on troops' ability to withstand long periods of intense combat with limited resupply and rest. A significant Russian failure was their inability to conduct effective small unit infantry operations in a MOUT environment. While the Russians captured Grozny in January 2000, the city is now unlivable due to the massive destruction caused by Russian artillery.

The battle of Grozny illustrates the type of urban insurgency that may dominate the 21st-century battlefield. Understanding the Chechen strategy, tactics, and weapons is essential in preparing U.S. forces for future urban battles.

Appendix D

THE URBAN AREA DURING STABILITY MISSIONS CASE STUDY: EAST TIMOR

LtCol Phil Gibbons
G3 Land Command, New Zealand Army

JOINT HEADQUARTERS
NEW ZEALAND FORCE
EAST TIMOR
(NZFOREM)
PRESENTATION
22 MAR 00

I am currently the G3 on the New Zealand Army's only operational level headquarters, Land Command. Land Command has been appointed the Joint Headquarters for the Joint NZ Force in East Timor (NZFOREM). I held the role of Chief of Staff for the Joint Headquarters for the entire planning period that led up to the deployment and during the deployment to East Timor.

On 21 October 1999, New Zealand deployed a battalion group to East Timor as part of the Australian-led International Force East Timor (INTERFET). New Zealand's contribution consisted of a Special Air Service element, an infantry battalion group supported by a troop of armored personnel carriers (APCs), six UH-1H Iroquois helicopters, two C-130 aircraft, two frigates, and a supply ship—a total of almost 1,500 defense force personnel. New Zealand's force, whilst exceptionally small by international standards, comprised approximately 15 percent of the total strength of the entire New Zealand defense force. On 23 February 2000, INTERFET transitioned to the UN peacekeeping force, United Nations Transition Force East Timor (UNTAET). As of March 2000, New Zealand's contribution to this peacekeeping force is an infantry battalion group, four UH-1H helicopters, and a supply ship.

N
Z
LAND COMMAND

N
Z
LAND COMMAND

N
Z
LAND COMMAND

N
Z
LAND COMMAND

N
Z
LAND COMMAND

I LOVE YOU
MILITARY
NEW ZEALAND
TIMOR LESTE
AMEU
LAND COMMAND

BRIEFING FORMAT

- NZDF Structure and Current Operations
- MOUT Operations, Doctrine and Training
- East Timor - the Environment
  - the Conflict
- Defining the MOUT problem
- Introduction to INTERFET, Mission, Intent, and Contributors
- Joint Planning Process
- Command and Control
- Overview of INTERFET Operations
- INTERFET / UNTAET transition
- UNTAET Mission, Intent and Contributors
- Lessons Learnt

The format for this presentation is shown on this slide. I am very aware that Dr. Russell Glenn has a fierce reputation for cutting speakers off who go over time so please bear with me as I move through the presentation at a reasonable pace.

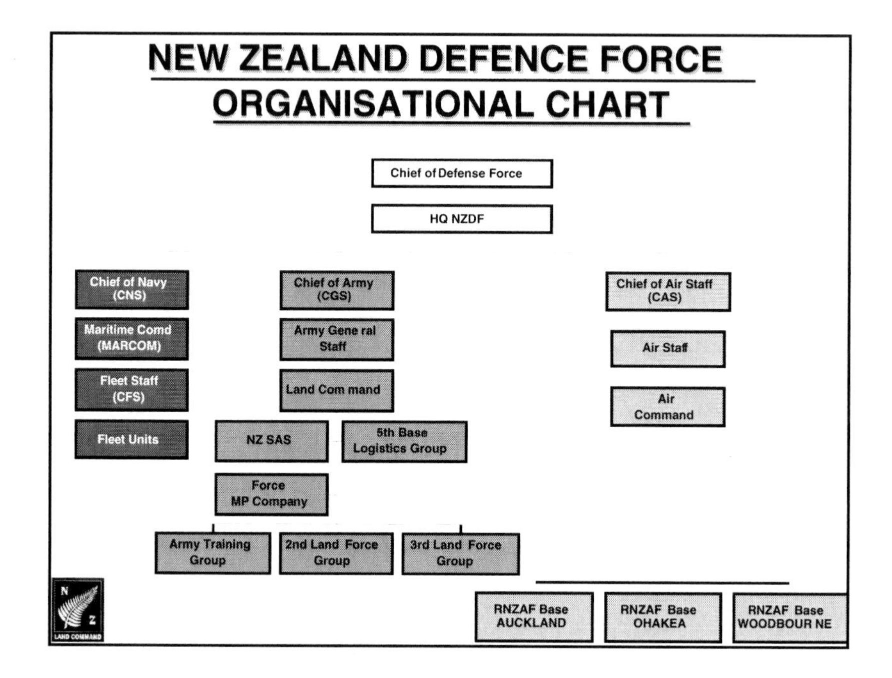

Before I commence the main part of the presentation I need you to understand the structure of the New Zealand Defense Force. There are three distinct services: Navy, Army, and Air Force.

The Navy is approximately 3,000 strong and based on four frigates, a supply ship, and several smaller specialist craft.

The Army is based on three formation-sized organisations and force troops of which our Special Forces, the SAS, are part. The Army Training Group is responsible for the conduct of all individual training from recruit through to the most senior of the officer courses. The 2nd Land Force Group contains the operationally deployable forces centred around a light battalion group. It is this formation that contributed the first contingent of New Zealanders to East Timor. The third formation is the 3rd Land Force Group that is responsible for the rotation forces. It is based on a Light Infantry Battalion but has limited combat and combat service support elements within the formation. The strength of the active army is 4,500. In addition to this figure there are an additional 2,500 reservists.

The Air Force is based around two Squadrons of A4 Skyhawks, two Boeing 727 strategic movers, five C-130 transports, four P3 Orions and 15 UH-1H Iroquois Helicopters. The Air Force is approximately 3,000 personnel strong.

I will discuss command and control with the New Zealand Defence Force later on in the presentation and relate it directly to East Timor.

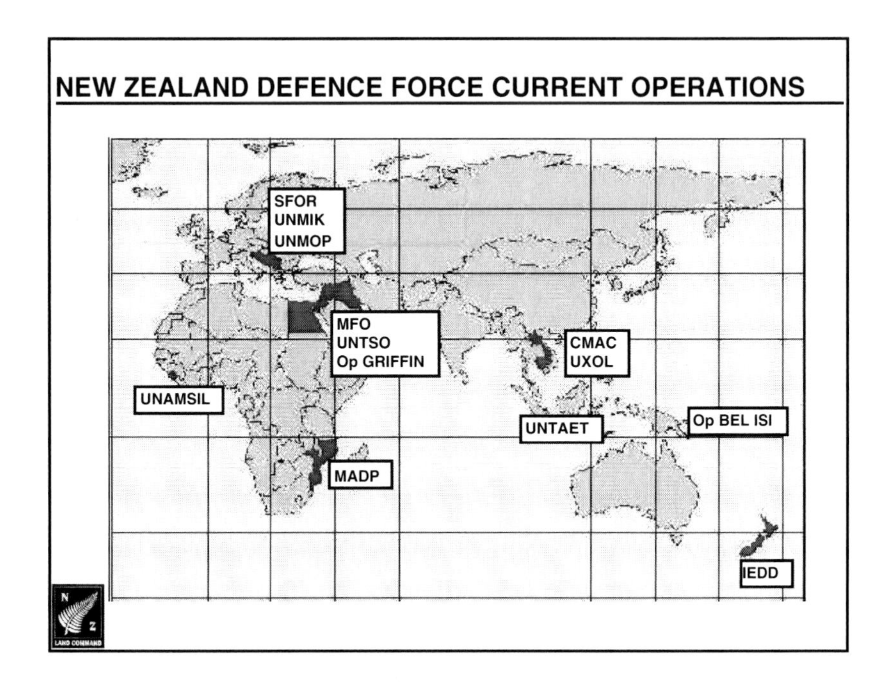

This slide displays the missions that the New Zealand Army are currently involved in. In January of this year the Army had 18 percent of its active forces deployed overseas on a variety of missions.

MOUT OPERATIONS

- HISTORY
- DOCTRINE
- TRAINING
- EAST TIMOR

For the next phase of the presentation I will very briefly address the issues outlined on this slide.

HISTORY OF NZ ARMY MOUT OPERATIONS

Significant Experience During Major Conflicts

'The New Zealand soldier is physically fit and strong. He is well-trained and formidable in close-range fighting and steadier than an Englishman. He does not shrink from hand-to-hand fighting. In many cases strong points had to be wiped out to the last man, as they refused to surrender.'

(German 14 Panzer Corps Post Operation Report after Battle for Cassino May 1944)

World War II was the last major conflict that saw New Zealand forces exposed to conventional MOUT operations.

In that instance the NZ Army proved to be very capable and effective in this type of operation. This level of competence is represented in the quote on this slide.

Since that conflict the NZ Army has been committed to Korea, Malaya, the Indonesian Confrontation, Vietnam, and several peacekeeping missions. None, however, have provided the MOUT experience such as that from World War II.

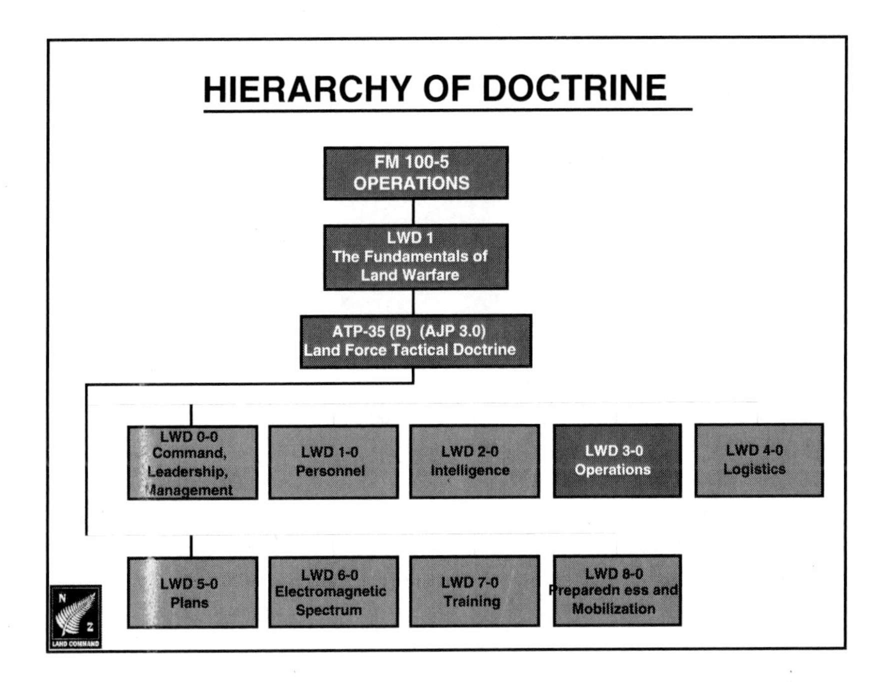

Given the limitations on the resources available to the New Zealand Army, we are not in a position to produce our own doctrine in any form. We are completely reliant on the doctrine of other nations. The capstone document for our doctrine is the U.S. FM 100-5, *Operations*. We combine this key document with the Australian Manual of Land Warfare Doctrine series. This slide demonstrates the flow of doctrine through the critical path to the Land Warfare Doctrine 3-0, *Operations*.

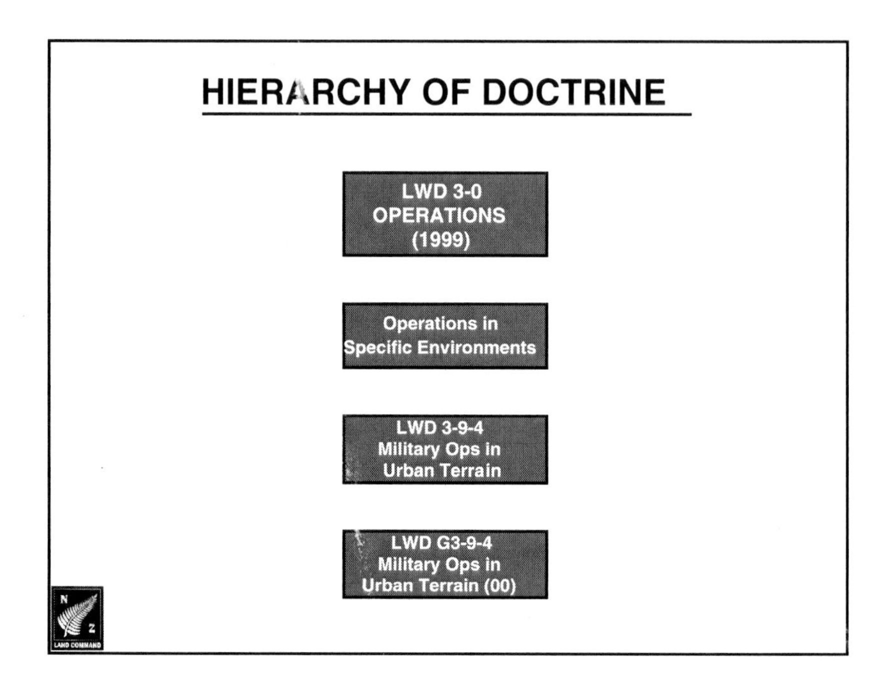

This slide follows on from the previous slide and leads from Land Warfare Doctrine 3-0, *Operations*, into the tactical level. At the tactical level, the primacy for MOUT doctrine for the New Zealand Army lies with the Australian Defence Force. Australian MOUT doctrine has had a gestation period of at least 16 years. Shortly, however, Australia-produced MOUT Doctrine is expected to emerge as developing doctrine. The two MOUT documents shown on this slide are expected to be on line this year. When they do, the New Zealand Army will formally adopt these references.

MOUT TRAINING

- Modest Facilities
- Reliance on '3rd Party'
- Mission Specific
- Survival in Built-Up Areas

Between World War II and the latest of our peace support operations, the New Zealand Army has had little experience with MOUT. From Korea through Malaya, the Indonesian confrontation and Vietnam, the New Zealand Army has developed considerable expertise in light infantry operations in the jungle environment. MOUT has been and to a degree still remains an aberration. Whilst the need for MOUT training has been recognized in post operational studies, it has never been translated into army-wide policies. Efforts by individuals and some units have improved matters. For example, this approach has resulted in the commitment to a very modest MOUT facility being constructed in our major training area. Whilst physically modest, it does demonstrate a degree of commitment to developing the skills associated with this special operation.

In recent times, when we have deployed forces to a potential MOUT environment, we have had to rely on third party assistance for the provision of training facilities. For example, the force that we first deployed to Bosnia in 1994 completed a month's training in the United Kingdom at the Copehill Down facility. Where possible, our forces that deployed to East Timor used the MOUT facilities in Townsville in Australia en route to East Timor. As result of this reliance on third party facilities, the majority of our training has to be mission specific and emphasizes force protection within a UN peace support operating environment rather than offensive action in a conventional MOUT environment.

EAST TIMOR

- The Environment
- The Conflict
- Defining the MOUT Problem

The next few slides have the purpose of introducing to you East Timor and in particular:

- The environment of East Timor as it confronted the International Force East Timor (INTERFET)
- The background to the conflict
- Defining the MOUT problem

I should add at this point that East Timor is some 5,000 km from New Zealand. To put that in perspective, it is a 10-hour flight in a C-130 transport aircraft. As you can imagine, this provided us with a considerable strategic move problem.

THE ENVIRONMENT

Timor has a rugged mountainous spine running east-west down its length. The highest point is in excess of 10,000 feet. The rest of the island drops away to either side of this range towards the coast in the north and south. The coastal plains are generally narrow with the northern coast being more densely populated. The rugged terrain on East Timor is a constraint to operational mobility, especially in the mountainous interior where vehicle movement is limited to the very few roads available.

East Timor has two annual seasons determined by the monsoon regime. The wet monsoons come from the NW carrying heavy rains during the period November through April of each year. The dry season turns the northern coast into an arid landscape whilst the mountainous and southern zones generally remain quite green. Temperatures average around 30 degrees Celsius.

Indonesia built an extensive road network, mainly for military purposes. However, the roads are generally poorly maintained and are in many places impassable during the wet season. A sealed road runs parallel to the northeastern coastline linking the main population centres of Dili and Bacau. Lateral routes in the north and all main routes in the south are one lane but are for the most part sealed.

N
Z
LAND COMMAND

N
Z
LAND COMMAND

N
Z
LAND COMMAND

N
Z
LAND COMMAND

N
Z
LAND COMMAND

N
Z
LAND COMMAND

N
Z
LAND COMMAND

THE CONFLICT

East Timor was a Portuguese possession until early 1975. In the power vacuum created by the departure of Portugal, one of the major native political parties gained control of much of the territory after a brief fight against internal opposition parties. It declared independence in November 1975 as the Democratic Republic of East Timor. The area was subsequently invaded and occupied by Indonesian forces in December 1975. In 1976 Indonesia declared that East Timor was its 27th province. Although the guerrilla movement continued to resist annexation throughout the 1980s and 1990s, Indonesia faced little international opposition to its policies on the island. The UN did, however, dispute the legality of Indonesia's administration in East Timor.

In July 1998, President Habibie of Indonesia offered East Timor special autonomy. This turned into a landmark decision to allow East Timor the chance to gain independence by voting on its future. On 30 August 1999, over 98 percent of the East Timorese came out to vote; 78.8 percent voted in favour of independence. This result was marred when pro-integrationist militias went on the rampage, looting, destroying buildings, killing many, and forcing others to hide in the interior. It has been alleged that the Indonesian Armed Forces were covertly supporting these militia elements. This in turn prompted the UN observer force on the island to withdraw from East Timor. Reports of atrocities and the forced exodus of East Timorese to West Timor eventually led to the deployment of INTERFET in September 1999.

TOTAL

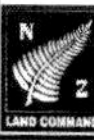

54

N
Z
LAND COMMAND

N
Z
LAND COMMAND

N
Z
LAND COMMAND

DEFINING THE MOUT PROBLEM

During the planning phase of this mission, INTERFET considered the security and control of the border between East and West Timor its main effort. Additionally, it was determined that within East Timor the issue was one of lawlessness rather than security. In the west it was more of a low-level insurgency problem. West Timor had few populated areas and the majority of the dwellings had been destroyed.

The threat was further defined as the planning progressed. This resulted in the identification of several generic tasks:

- Establishment of a secure environment in East Timor
- Protection of key points such as the airport and port facilities
- Separating Indonesian forces from the general populace

However, in the first phase of the operation the greatest threat was expected to be found in the island's primary urban areas, Dili and Bacau. This concern was certainly supported by the media and other reports that were originating from East Timor in August and September of 1999. Both cities had populations in excess of 100,000 people.

By the time the operation commenced in October 1999, the major population areas in the west had been largely deserted. Either the militia had forced the people to leave or the Indonesian Armed Forces had evacuated them to West Timor. Dili, indeed was all but deserted when the first INTERFET forces arrived in the city. People were in some cases still being forcibly evacuated by the Indonesians both by air and sea. Militia were present on the island, but their influence was limited. Their former harassment and killing of East Timorese had by this time been reduced dramatically. The priority tasks for INTERFET were:

- Deployment.
- Establishing a secure environment in the greater Dili area.
- Separating the Indonesian armed forces from the populace and the military's subsequent removal.
- Expansion operations beyond Dili.
- Securing of the east/west border.

LAND COMMAND

INTERFET

(INTERNATIONAL FORCE EAST TIMOR)

Concept of Operations

I will now introduce the INTERFET concept of operations through an indicative force mission and commander's intent.

INTERFET MISSION

To contribute to the maintenance of a secure environment in East Timor so as to facilitate humanitarian assistance and allow for the establishment of a UN PKF.

This slides shows a mission statement similar to that adopted by INTERFET.

COMD INTERFET INTENT

To establish a multi-national military presence in East Timor in order to conduct operations that will allow for the re-establishing of a stable and secure environment whereby the UN can assume command of the mission.

This slide shows a commander's intent akin to that developed by General Cosgrove, the Australian officer who served as INTERFET Force Commander. The photograph shows General Cosgrove addressing NZ troops in Dili shortly after the operation commenced.

CONTRIBUTING NATIONS

Australia	Nepal
Brazil	New Zealand
Brunei	Portugal
Canada	Philippines
Fiji	Singapore
France	Thailand
Ireland	United Kingdom
Malaysia	United States

These are the countries that contributed forces to INTERFET.

CDF NZ INTENT

To deploy NZDF FE as part of INTERFET to conduct operations in East Timor in support of Strategic National Objectives as determined by the Government of New Zealand.

This slide shows the New Zealand Chief of Defense Force's intent that was derived from the guidance provided by the Government of New Zealand.

NZ JOINT COMD MISSION

To mobilise, prepare, deploy, and sustain NZFOREM in order to achieve the NZ Military Endstate

This is the New Zealand Joint Commander's mission.

JOINT PLANNING PROCESS

Conducted in three phases

- Services Protected Evacuation Contingency (Company Group)
- Company Group, Naval Assets, Fixed Wing and Helicopter
- Battalion Group, Naval Assets, Fixed Wing and Helicopter

Planning within the New Zealand Defense Force took place in three distinct phases and resulted in three different force options. The first phase occurred concurrently with the deterioration of the situation in East Timor in May/June 1999. The second and third phases took place once the New Zealand government had determined that it was going to commit forces to the coalition INTERFET. The major difference between Phases Two and Three can best be assessed in terms of capability. The first and second involved small task-orientated forces designed to complement a larger force. The ability for independent operations by these force groupings was very limited. The third option presented a larger and potentially more employable force that was clearly able to undertake independent operations. The major capabilities contained within this third option force were:

- One light infantry battalion
- Medical support
- Organic combat service support

- Mobility through organic helicopters and wheeled and tracked vehicles
- Fixed wing transport aircraft, rotary wing aircraft, special forces, frigates, and a supply ship.

PLANNING ISSUES

Lack of Operational Task Visibility (MOUT?)

Deployment

M113/UH-1H Helicopter Refurbishment

Japanese Encephalitis Vaccinations

Purchase of Non-Military Vehicles

Perhaps the biggest issue facing us during the planning of this operation was a lack of what we call task visibility. Whilst we were able to gather information on the conflict and the environment, there was real difficulty determining the courses of action that would be pursued by the Indonesian Forces and the militia. Would they attempt to deny the INTERFET forces the opportunity to secure the urban areas of Dili as did the Chechens deny access to Grozny? Would they simply harass the security forces as we have seen in Northern Ireland? Until the most likely and most dangerous course had been established it was difficult to identify suitable force structures and specific tasks.

The strategic deployment of any sizeable force was going to cause a problem for the New Zealand Defense Force. Strategic movement assets are very limited. We eventually resolved this through the use of civil resources; third party assistance was an option, but one not employed.

Another significant requirement for deployment was the need to refurbish the M113 armored vehicles that would be needed for the deployment. There was also a UH-1H helicopter refurbishment requirement. The issue with both major systems was one of upgrading rather than introducing any new capability.

Japanese Encephalitis vaccinations became a critical issue when it was discovered that this vaccination required a 38-day period before it would become effective. Early on it became the critical path for our deployment.

At the time of deployment the Landrover was about to be replaced as our light vehicle by a nonmilitary variant. The Landrover was considered nondeployable. The nonmilitary vehicle deployed is shown on this slide. Its procurement program had to be hastened considerably in order to meet the deployment timeline.

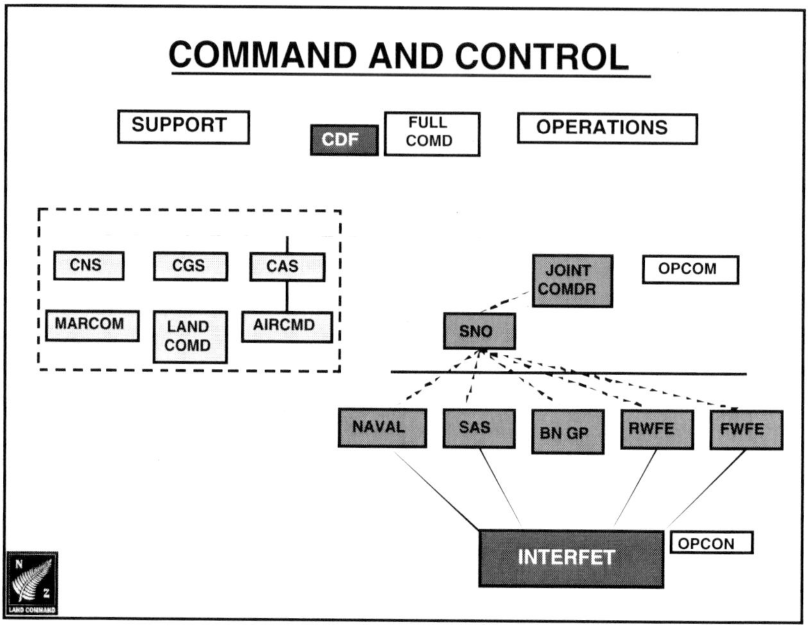

COMMAND AND CONTROL

SUPPORT
CDF
FULL COMD
OPERATIONS
CNS
CGS
CAS
MARCOM
LAND COMD
AIRCMD
JOINT COMDR
OPCOM
SNO
NAVAL
SAS
BN GP
RWFE
FWFE
INTERFET
OPCON
N Z
LAND COMMAND

This is the doctrinal C2 model that outlines both the operational and support command and control arrangements within the New Zealand Defense Force. The left-hand side of the model outlines the normal peacetime arrangements during which the Chief of Defense Force commands the three services through the respective service chiefs. The right hand side of the model demonstrates the command and control arrangements for an operational deployment using East Timor as the example. For East Timor, the Land Commander was appointed by the Chief of Defense Force as Joint Commander of New Zealand Force East Timor. The naval and air operational level commanders were directed to contribute force elements to the joint force; indeed, these elements were assigned to the operational command. Once the mobilization and strategic deployment of these force elements had been achieved, they were assigned by the New Zealand Joint Commander to the Commander INTERFET under operational control. The Senior National Officer is the senior New Zealand officer in theatre and represents New Zealand's interests on behalf of the New Zealand joint commander.

OVERVIEW - INTERFET OPS

Key Dates

13 Sep 99 Indonesia acceptance of UN Force

15 Sep 99 UN Resolution

20 Sep 99 INTERFET Deploy EAST TIMOR

I would now like to focus a little more on INTERFET operations. This slide shows the key dates leading up to the deployment of INTERFET.

MOUT OPERATIONS

- Military Operations (Chapter 7) to Restore a Secure Environment
- Specific Comd Estb for Control of Urban Area
- Comd and Planning Team Looked to Historical Examples for Lessons Learnt
- Importance of Mission Analysis
- Allocation of Troops

As I have already explained, the primary objective for INTERFET was to restore a secure environment with the priority on the urban area of Dili.

INTERFET achieved this in two very distinct phases. The first involved the securing of Dili by a task organized force consisting of Special Forces and light forces. Once this had been achieved, a task force was established to maintain security in Dili. This force remained intact throughout the entire INTERFET operation. It achieved considerable success given that it was a hastily derived coalition force.

Historical examples were examined early on in the planning process in order that previously learned lessons would not have to be relearned. Mogadishu was one of the examples that was identified as being relevant to the situation in East Timor.

The joint military appreciation process, and in particular the first step, mission analysis, was identified as the key to maximising the value drawn from the limited resources that were available to

INTERFET for solving the Dili security issue. It was crucial in correctly allocating tasks to the variety of force elements that made up the combined force called Dili Command. The command consisted of:

- an Italian paratrooper company
- an Australian mechanized battalion
- a New Zealand light infantry company
- an Australian airfield security squadron.

MOUT OPERATIONS (CONT)

Employment of SF

Use of Aviation Assets

Importance of Force Protection

Rotation of Forces

Well Balanced Reserve

Route Control

There are real opportunities available for SF during a Chapter 7 operation. Early on during the INTERFET deployment this force element facilitated pinpoint targeting and effective building clearance.

There were helicopters airborne 24 hours a day throughout the initial securing of Dili in order to provide constant observation and reaction. This proved to be a very successful concept.

During the entire operation, force protection was paramount. There was significant pressure at the political and strategic levels to ensure that all force protection measures were maximized. It remained a key factor in all planning and during the subsequent conduct of operations.

It was essential to continually rotate forces during the operation to secure Dili due to the intensity of the operation. Commanders at all levels had to ensure that they had sufficient resources available to maintain 24-hour operations.

Whilst there were reserves maintained at all levels, there was a specific need to hold a well-balanced reserve centrally at brigade level. This reserve had to meet the need for a balanced force in both overall capability and deployability.

Route control became a critical requirement as the operation to secure Dili progressed. As more and more forces were committed to the operation, it became evident that a detailed plan for route control was required in order to avoid confusion and fratricide.

INTERFET / UNTAET TRANSITION

I would now like to briefly look at the INTERFET/UNTAET transition.

KEY DATES

1 Feb 00	Transition to UNTAET Commences
7 Feb 00	Sector East Transition
14 Feb 00	Sector Central Transition
21 Feb 00	Sector West Transition
23 Feb 00	INTERFET Transition Complete
Jul 00	Sec Gen Report to Sec Council
Mar 01	Current UNTAET Mandate Ends

This slide summarizes the key activities and dates in the transition process.

UNTAET MISSION

UNTAET conducts security operations and assists the SRSG to discharge the mandate in order to establish an environment of peace and security conducive to political, social, and economic development in East Timor.

The UNTAET mission was extracted from the UN Security Council resolution 1272.

COMD UNTAET INTENT

- Maintenance of security is the highest priority and the most important task for the PKF.
- The Chapter 7 mandate and ROE allow for a swift and robust response to any threat.
- Liaison with the major stakeholders is fundamental to ensuring security.
- Within force capabilities, PKF will assist with humanitarian and nation-building programs.

These were the key elements in the Commander UNTAET's intent.

UNTAET CONTRIBUTING NATIONS

- Australia
- Bangladesh
- Brazil
- Canada
- Chile
- Denmark
- Egypt
- Fiji
- France
- Ireland
- Italy
- Jordan
- Kenya
- Nepal
- New Zealand
- Norway
- Republic of Korea
- Pakistan
- Philippines
- Portugal
- Russia
- Singapore
- Thailand
- United Kingdom

UNTAET CONCEPT OF OPERATIONS

NZ DEFENCE FORCE
LESSONS LEARNT

Political

Strategic

Operational

Tactical

I would like to conclude the presentation by highlighting some of the key lessons learnt at the various levels indicated on this slide.

POLITICAL LESSONS LEARNT

- Positive and high public profile
- Real political will in New Zealand for defense force to be heavily involved in this type of operation

Throughout all phases of the mission there was an incredible level of internal public support for the New Zealand forces. This was mainly attributable to highly positive media support for New Zealand's participation.

Considerable public pressure was brought to bear on the government of New Zealand to contribute positively to the security situation in East Timor. The end result was unanimous political support for the deployment of New Zealand forces to East Timor.

STRATEGIC LESSONS LEARNT

- Logistic support coordination very difficult for lead nation
- Coalition partners are expected to provide a reasonable degree of self-sufficiency
- Involve political decision makers early
- Understanding response times

Lead nation status was assumed by Australia, the senior partner in the INTERFET coalition force. Australia found it difficult to provide the logistic support required by the entire force as contributing nations arrived in theatre with various levels of self sufficiency. The operation proved to the logisticians involved that there is no longer the redundancy there once was in the days of mass logistics.

In any form of coalition operation, contributing nations are expected to contribute a complete capability and the self sufficiency required to sustain it. Combined capabilities are considered the exception but it is often achievable in the areas of combat service support and health support.

This deployment clearly indicated the utility of involving political decision makers early in the process. This engagement resulted in timely decision making and allowed for a deliberate process during the preparation of force elements.

This operation highlighted the continued requirement to ensure that politicians understand the importance of response times and their effects. Response times are established so that resource allocation can be managed and mission essential training activities completed according to schedule.

OPERATIONAL LESSONS LEARNT

- Use of doctrinal basis for identification of possible tactical tasks (MOUT)
- Maximise combined planning opportunities
- Shape the MAP, use the process, identify the factors
- Previous msn experience—Planning/Implementation

It is essential that when there is little to no information available on what specific tasks will be completed during the early stages of the operation, the force elements should prepare by training to the doctrine based mission essential tasks. This enables a force to achieve the required level of capability within the assigned response time and achieve concurrent activity through planning and force preparation.

Coalition operations provide considerable scope for maximising the opportunities for the development of combined planning procedures and doctrine. This was effectively achieved between Australia and New Zealand.

The military appreciation process is a tool of utility in the planning for any operation. It does not however need to be slavishly followed; instead it should be shaped so as to allow for the analysis of all the influences that may impact on an operation.

For the New Zealand Defense Force, the previous mission experience contained within the planning teams and force elements alike has been considered essential. Without this key ingredient the planning and conduct of this operation would have been significantly more difficult. We were able to benefit from the lessons learnt from Bosnia.

TACTICAL LESSONS LEARNT

- Peace support operations require MOUT skills
- Interoperability with AS
- Response forces train to all METLs

The East Timor operation has reinforced the need for deployable forces to be familiar with the essential elements of MOUT operations. Whilst New Zealand forces have not been involved in a conventional MOUT operation since World War II, peace support operations such as East Timor and Bosnia have proven that MOUT core skills are required in forces expected to deploy rapidly to undertake peace support operations.

Interoperability with Australia was an issue very early in the INTERFET operation. It had been two and a half years since New Zealand and Australia had exercised together in a combined environment at a level greater than battalion group. As the operation progressed it obviously became a lesser problem, but early on it was an operational risk.

The issue of doctrine and training has already been touched on; it is one that warrants repeating. If response forces are to minimize operational risk they must have trained to an established list of mission essential task standards that are doctrine based.

Appendix E

FORCE PROTECTION DURING URBAN OPERATIONS CASE STUDY: MOGADISHU

Urban Force Protection

LtCol John Allison, U.S. Marine Corps (ret.)

Force Protection During Urban Operations

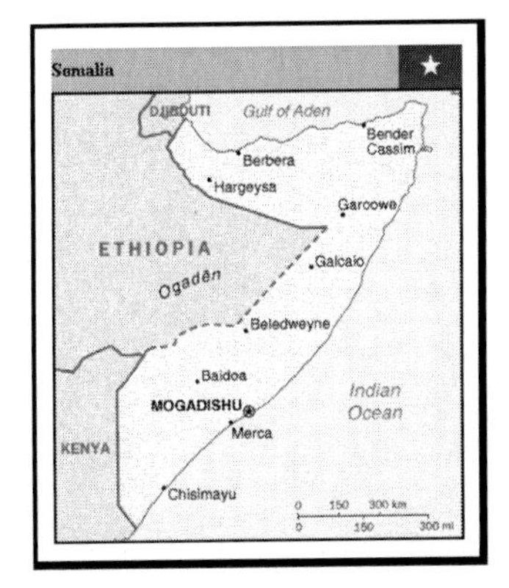

Mr. John Allison, Director of Program Integration, MCWL

Images with © are copyrighted by Philadelphia Enquirer

Photos taken by Pete Tobia

Thank you for the opportunity to discuss my experiences in Somalia. I am going to talk about how we in the Marine Corps conducted one aspect of force protection. I was the leader of a team that was put together by our commander at that time, Major General Wilhelm, now Commander in Chief of Southern Command. My team and I were tasked to go out and look at our force protection posture by conducting security assessments throughout Somalia.

BACKGROUND

1988	Civil War Breaks Out
Jan 1991	Operation Eastern Exit (NEO)
May 1992	Siad Barre's Defeat & Exile
Aug 1992	Operation Provide Relief
Fall 1992	UN Battalion Deployed to Mogadishu
Dec 1992	Operation Restore Hope
Apr 1993	Operation Continue Hope (UNOSOM II)
Jan 1995	Operation United Shield

Here you see the timeline of events that led to the United States' involvement in Somalia. I'm not going to get into detail about them. We have other speakers on the schedule, such as Ambassador Oakley, who will talk about events at the strategic and operational levels. As background you can see that Marine forces got involved in December 1992 as part of a larger force deployed under the auspices of United Nations Chapter 7 rules of engagement to restore order to a failed nation.

SITUATION

- **No central government, police, military, etc.**
- **Factions disputing each other for control**
- **Food and weapons were power**
- **Only distribution system working provided KHAT and arms**
- **Black market the only means to obtain anything**
- **Significant food stocks but no ability to deliver**
- **2 million people displaced and 1.5 million at risk of starvation**
- **Death rate high from starvation, violence, disease**
  - Mogadishu 150 + per day
  - Chisimayu 25 + per day
- **Banditry a way of life**
- **Weapons cheap, visible, present everywhere; necessary for survival**

The very complex situation was as shown here. There was no one from Somalia with whom we could plan. In fact, at this time the UN was hunkered down at the airport because they couldn't move. There was a small UN battalion that was trying to help move food supplies that was finally overwhelmed by the local gangs and warring clans. Power was held by those who held food and weapons. They were the coin of the realm. The death rate was very high; it was estimated at 150 per day in Mogadishu and 25 plus per day in Chisimayu, the next largest community in southern Somalia. It is believed that a majority of those deaths were not famine-related, but rather the result of violence. Banditry had become a way of life. We were about to come in and take that away from them. Consider the implications of that. Cheap weapons were everywhere. They were needed for survival by the people, who had their own neighborhood watch program against gangs and bandits. Weapons were also what the gangs, bandits, and warlords used to suppress the remainder of the people. We used the weapons as a measure of effectiveness, i.e., when they went from being cheap and visible to expensive and hard to get we knew we were being very effective.

“Technical” vehicles are basically any type of Toyota vehicle or Jeep with a large caliber weapon system. These systems ranged up to 106mm recoilless rifles. Anyone with experience with that weapon in Vietnam knows how devastating it can be. The term “technical” came from NGOs and PVOs—non-governmental organizations and private voluntary organizations. When the situation deteriorated, these organizations had to hire gangs or clan members to protect their convoys. It was a form of extortion. They then sent vouchers back to their headquarters; the cost of this protection was noted as “technical assistance.” That is how they got the money to pay these thugs for protection when they moved supplies from ports and airfields to locations where people were actually starving. Another picture here shows a Somali in the shadows. He could be a sniper; he could be a bandit; or he could be protecting his home. How could you know? These types of things were part of that complex environment.

The picture at the bottom center shows the types of weapons we confiscated when we first landed. These are what we faced when we went ashore in December 1992: artillery pieces, armored vehicles, and tanks. The picture of Aideed represents the political complexities of a civil war. Aideed, Ali Mahdi, and their organizations (such as the United Somali Congress and Somali National Alliance) created very complex factional dynamics.

You talk about the fourth estate; the photograph at the top right shows a little real estate belonging to the fourth estate. It is a press camp—actually this picture came from United Shield, but it reflects the volume of press representatives that came to Somalia. We had them with us throughout the country at the time. In fact, you may remember that media influence was one of the factors that hastened deployment and caused us to become involved in a lot of the other local and international "political" issues. So media influence has to be factored into your planning. And then there was the UN—the lower right is a picture of UN vehicles at the airport. When we landed in December 1992, that is where they were hunkered down. They had established a perimeter so that we could get aircraft in and get a look at what kind of situation we were facing.

These pictures show the local population and the types of conditions that we had to deal with. Are these just young men that happen to be in an area affected by the civil war, or are they part of a warlord's faction? Are they gang members? Bandits? We didn't know. Violence was everywhere. The photo at the bottom left shows a Somali who had been hit by an AK-47 round and was being treated in very austere conditions. There is a picture of the civil-military operations center (CMOC). We had a large consortium of NGOs and PVOs with which we worked out missions for the next day's work in the humanitarian relief sectors and planned the movement of the many relief convoys. It was the way that we interfaced with the UNITAF headquarters, a joint and combined task force that included all of the NGOs and PVOs. In the city of Mogadishu the infrastructure had completely broken down; there was no power or water. It was resultantly a very primitive urban environment. The center picture shows the core element of our initial mission: taking care of starving people. It is a picture of a feeding tent located in the hinterland's Bardera famine triangle.

A FAILED NATION MISSION

Secure major air and sea ports, key installations and food distribution points; provide for open and free passage of relief supplies; provide security for convoys and relief organization operations and assist UN/NGOs in providing humanitarian relief under UN auspices.

UNITAF SOMALIA Dec 1992

This is a failed nation mission. It is the mission that we were given as the UNITAF Marine component (Marine Forces Somalia, MFS). You can see how difficult this mission might be given its many complexities.

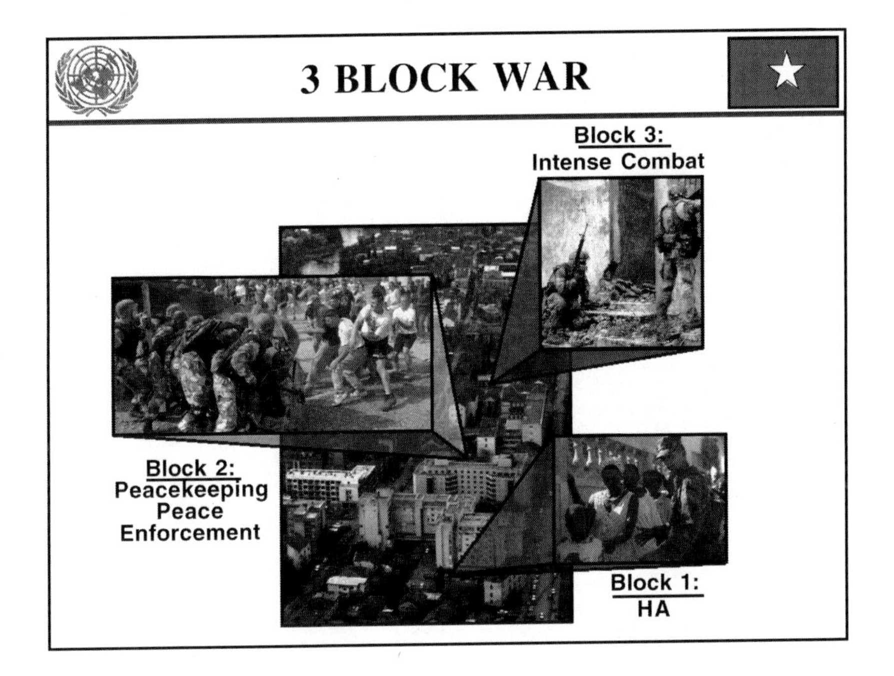

This is the situation in which the Marine Corps found itself, though we had not coined the term "three-block war." Our former commandant, General Krulak, later introduced it. We found ourselves smack dab in the middle of a three-block war. At one moment we would be doing a humanitarian assistance mission. The next moment, we were separating warring clans in another part of the city while continuing that humanitarian assistance. And there were occasions when we were involved in intense combat in another part of the city while remaining committed to the previous two missions. I saw such a situation unfold in January 1993. All three missions took place simultaneously in Mogadishu. At the same time, our headquarters was responsible for other areas in Somalia that were assigned to the Marine component.

FORCE PROTECTION

"Security program designed to protect service members, civilian employees, family members, facilities, and equipment, in all locations and situations, accomplished through planned and integrated application combating terrorism, physical security, operations security, personal protective services, and supported by intelligence, counterintelligence, and other security programs." ***JP 1-02***

OP 6: Provide Operational Protection.

...this task includes protecting joint and multinational air, space, land, sea, and special operations forces; bases; and LOCs from enemy operational maneuver and concentrated enemy air, ground, and sea attack; natural occurrences; and terrorist attack. This task also pertains to protection of operational level forces, systems, and civil infrastructure of friendly nations and groups in military operations other than war.

UNTL 1996

Force protection. This was not yet a term really familiar to us in 1992 and 1993. In fact, it has only been in the last few years that we even put the term in a joint doctrinal publication. The first paragraph provides the definition of force protection for you. We also have universal joint task lists that have evolved over the last few years. They encompass what joint task force commanders and commanders-in-chief expect their service components to be able to execute. For the Navy and Marine Corps we have what's called the Universal Naval Task List (UNTL). We put these tasks together in 1996 using lessons learned from our operations. The second paragraph is an extract from an operational level force protection task. It pertains to the protection of operational forces, friendly nations' infrastructure, and other groups during military operations other than war.

FORCE PROTECTION

OP 6.5: Provide Security for Operational Forces and Means

To enhance freedom of action by identifying and reducing friendly vulnerability to hostile acts, influence, or surprise. This includes measures to protect from surprise, observation, detection, interference, espionage, terrorism, and sabotage. This task includes actions for protecting and securing the flanks and rear area of operational formations, and protecting and securing critical installations, facilities, systems, and air, land, and sea LOCs.

Universal Naval Task List

Drilling down further into this task we see "Provide security for operational forces and means." I just want to call your attention to the elements shown. They are measures taken to protect organizations from surprise, observation, detection, interference, espionage, terrorism, and sabotage. We want to reduce friendly vulnerability. These are formal elements that have been derived from our lessons learned.

Now we are going to talk about what we did in Somalia before we called it "force protection."

These pictures represent the elements that we employed in Somalia to achieve mission success. A previous picture showed the group of people seen at the upper left. We didn't know who they were or what they were doing. But now you see a Somali interpreter with a Marine patrol, a counter-intelligence team. They were out among the population, moving, talking, and discussing things. It was a very important aspect of how we got our intelligence and developed situational awareness.

The center top picture shows a strong point security post; I call them Ft. Apaches. Anyone who's familiar with our American folklore and culture from the west knows we had forts throughout the western states and territories during our nation's expansion. These forts provided safe havens for Americans as they moved west. As we moved into the urban environment, Task Force Mogadishu, at that time commanded by then Colonel (now Major General) Bedard, was establishing such strong points throughout our sector of Mogadishu. So we had these little Ft. Apaches out there where the friendly civilian folks could come in and be protected. At the same time we had safe havens for our own people to use when they moved through the city. Then we had engineers and elements of the MARFOR Somalia naval construction regiment conduct Operation Clean Street. They went out and literally cleaned streets. We removed rubble, restored water supplies where we could, did the same with power, and brought amenities to the civilian population in order to win them over to our team, to show them that we were there to help. We also provided medical and dental support. As we did so, we became more familiar with the local population. Marines learned more about their culture and then they started joining our team. As a result, we had better situational awareness. We also helped stand up local markets, a significant measure of effectiveness. If the market was flourishing, things were starting to get back to normal. If no one was going to the market and it wasn't a religious holiday, we knew we had something to worry about. It was a key indicator. The next thing we did was conduct a psychological campaign. We established a local paper, distributed leaflets, and sent messages using vehicles equipped with speakers. When we found a compound that held a gang or warlord element we would go in with PSYOP assets and tell them that if they didn't come out and put their weapons down within 15 minutes, we were coming in to get them. We backed it up by force. And that's a critical point: you must back up PSYOP and non-lethal weapons with lethal force.

Rules of engagement. I believe to this day that one of the reasons we had fewer casualties and great mission success during Operation Restore Hope is that we went in with very good ROE under Chapter 7. If a Marine felt that he was threatened by a Somali who had a weapon and would not put it down, the Marine could engage. That was the essence of it. If you felt threatened, you could shoot. You didn't have to wait to be shot at to return fire. Later in the brief I'll show you what kind of impact that had psychologically on the Somalis. It was a key element.

SECURITY ASSESSMENT

- **Vulnerability Assessment Team**
- **OIC: Field Grade Officer (Combat Arms)**
- **Intelligence Officer: (Fusion)**
  - **CIT / HUMINT, SIGINT, TIO**
- **Force Recon: Infiltration Specialist**
- **Military Police: Officer/SNCO**
- **Combat Camera**
- **Engineer**
- **Other (i.e., CST & PSYOP)**

As we set all of this into motion and started executing our missions, General Wilhelm wanted to start a security assessment team. Security had to be continuous: 24 hours a day, 7 days a week. Now some of what is in this slide may seem very rudimentary to those of you in the military profession. However, when you get into these complex situations, oftentimes security falls to the bottom because the force is so busy dealing with the press and other political aspects of the operation that there is tendency to look beyond what's going on in the security environment. General Wilhelm therefore formed an assessment team. He put a field grade, combat arms officer in charge; that was my role. We also had an intelligence officer to fuse Counter Intelligence (CIT), Human Intelligence (HUMINT), and SIGINT, act as Target Information Officer (TIO), and tell us what's going on in our environment. The intelligence summary came in every day. We used specialists from force reconnaissance. I had an infiltration specialist, a staff sergeant, who completed very detailed drawings of all our positions from the threat's perspective in order to see where we were vulnerable. We could then plan responses to anticipated threats. The military police were also invaluable. When there are no police in a country, the military police have an understanding of physical security and criminal activity that further reveals areas in which a force is vulnerable. They provide a perspective on dangers that may not be evident from a straightforward military point of view. Combat camera was another valuable asset. You need to be able to take pictures so that commanders and their staffs at various levels can understand the situation because they can't get out to all significant locations. We used digital video from which we could make pictures to print as part of our reports. The team included an engineer officer. He needed to look at the situation with respect to barriers, lighting, and what kinds of things we could do to enhance the position. He made estimates of the types of materials we would need. Other team members included a coalition support team (CST) member, someone who could interpret. I went out to Baledogle, where we had our MFS aviation element and where we had Tunisians providing security around the perimeter. Well, I would go out with my team and ask questions of sentries, just to get a feel for their awareness, to see what's going on and determine whether they needed anything. The sentry could speak three languages, none of which was English.

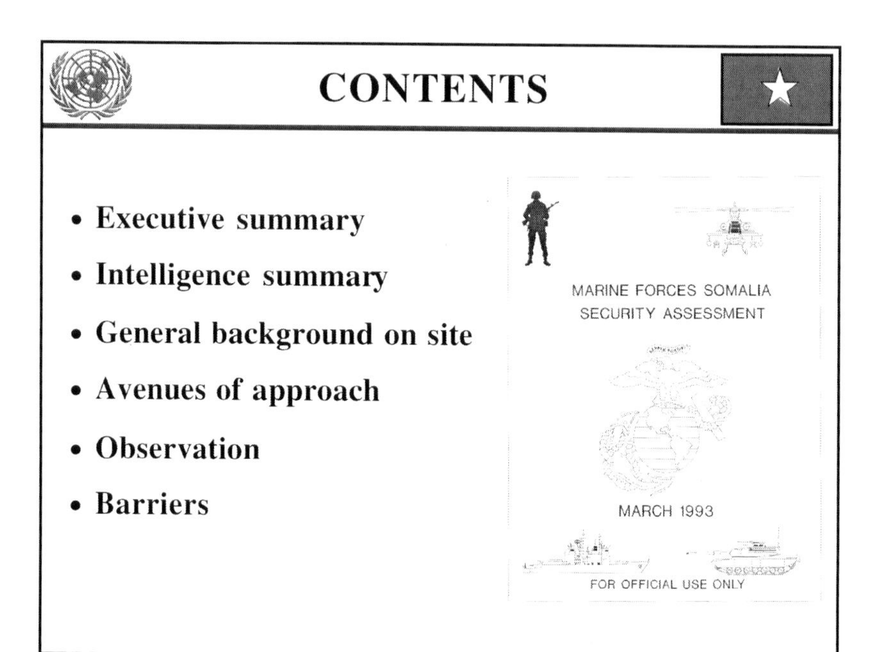

Here are the key elements of the *Marine Forces Somalia Security Assessment*. This report had information pertinent to both U.S. forces and our coalition partners. However, there were limits on what we could and could not share based on policy and laws. For each location there was general background on the site: why the site was important, what it was being used for, its role in the mission, and so on and so forth. Avenues of approach: How do you get to the site? Which roadways are a concern? Which waterways? If the threat has some type of aviation, what are the ways that they could come in from the air? Observation. A key here was not only what you could see, but what the enemy could see. We would go outside the compound with an infiltration specialist and look at our positions. One weak area that we found was our airfield. We took rounds in some of our Cobra helicopters while they were sitting at the airfield. We found that there was an old factory building overlooking the airfield from which a sniper could hit the aircraft. We didn't have enough people to maintain a watch over every possible position from which an enemy could engage the airfield. What we did was take a lot of sea cargo containers; anyone who went to Somalia may have seen them around the airfield. We stacked them as high as we could, six high, and put them around the airplanes and helicopters to block a sniper's field of fire. We grabbed what was on hand and it worked. Constructing protective barriers was another challenge. We all remember how a truck slammed into our Beirut compound. Since then our installations have different types of barriers that force drivers through a series of S curves. You may not have that material at hand when arriving in some nations. Again, sea cargo containers might be one solution.

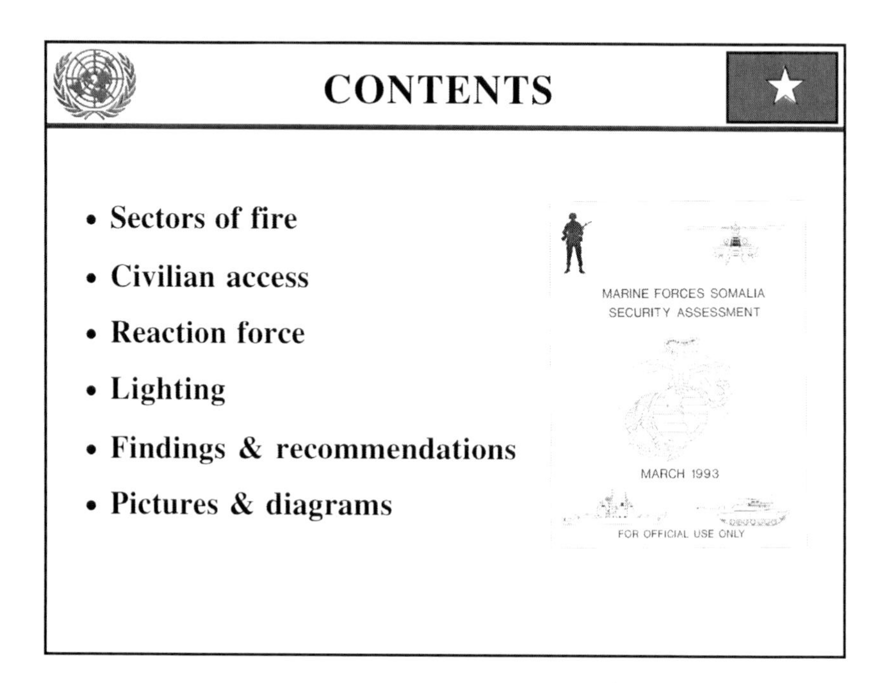

Sectors of fire were a big issue. In an urban environment we all know that you may have a unit from another service or country within 200–300 meters of your positions. In Mogadishu we had other countries working with us. So we had language problems. In addition, we had army forces that were supporting the Joint Logistics Command in Mogadishu and other parts of the country. They would set up defenses and we might not be aware of their fire plan. The next thing we knew, we had units with weapons pointed at other friendly forces. These are the sorts of things that you have to take into consideration.

Civilian access—that's a big one. One thing you do when you establish a compound is determine who is to have access. Oftentimes you're trying to get the economy going so you're hiring local civilians to work in your compound. Who are these civilians and what is their background? Do you have an access roster? Do you have a badge and ID system so you know who's going in and out?

A reaction force—always know who your reaction force is, how you can get hold of them, and what their reaction time is. You need to exercise the reaction force. You can't just say "I've got one" and when you call on them find that their radio net has been changed, that they cannot be located, or that they're out doing another mission.

Lighting was another big issue over there. You're in an environment where there is no electrical grid, no power. You need to bring generators and lights. The more you can light your external perimeter so that your sentries can see, the better security you're going to have.

We found some perimeters that were lighting their own positions instead of the surrounding area. They weren't U.S. Marines, but we did find people who worked for us that were lighting their own positions. It made them great targets for snipers.

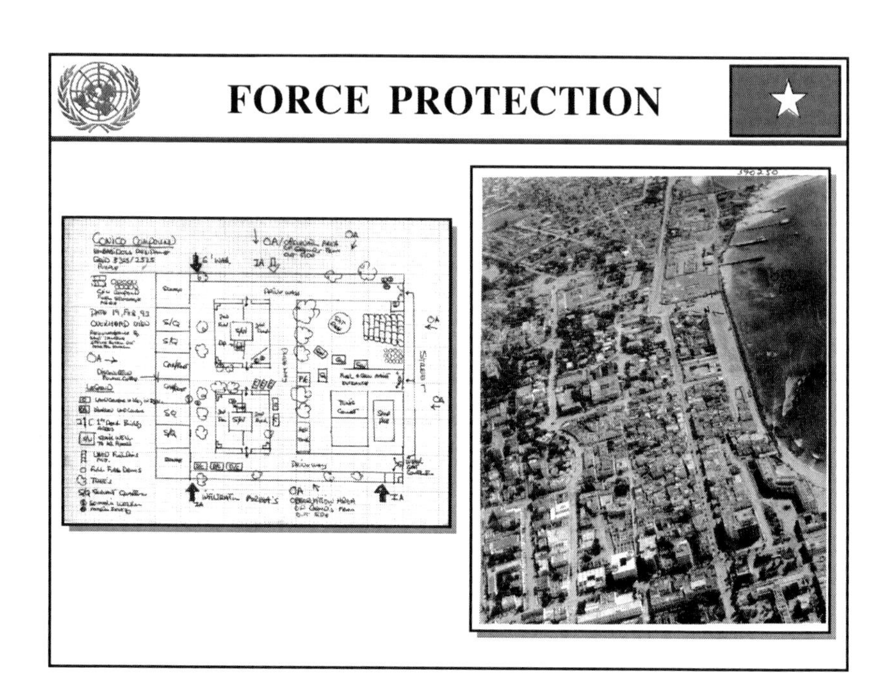

Here's the actual sketch regarding what I was telling you about earlier. Now, we're not known for being great spellers—this is the Conoco compound. Ambassador Oakley actually resided there while in Somalia. It was called the U.S. Liaison Office (USLO). We put Marines there to help with security and we looked at the vulnerability of his compound. The state department team and the ambassador were key to our mission's success. We needed to protect them. And we were no longer at our own embassy because that was occupied by Marine forces, joint and combined forces headquarters, and UNITAF headquarters.

Here is an aerial photograph of the Conoco compound. You need to take as many pictures as required to show your decisionmakers the situation at hand. It also helps you to see vulnerabilities from different perspectives. Take them from both the ground and the air. It is very easy to see urban avenues of approach and choke points in this manner.

I mentioned sectors of fire. This portrays the complexities we confronted. Pakistanis, Turks, Nigerians, Botswanans, Egyptians, and U.S. forces were all operating in Mogadishu. Mogadishu proper was actually divided between Marine Forces Somalia in the west and Italian Forces Somalia that had the older part of the city, the old port and eastern Mogadishu. The areas and the dividing line just so happened to fall on the boundary between Aideed's area (which is what we had) and Ali Mahdi's area (what the Italian forces had). We had to be aware of relevant complexities, the political considerations when we put our Ft. Apaches (strong security posts) around to quiet the town. One day for a 24-hour period we had no reports of anyone being killed in Mogadishu. It was a notable event; the first time since deployment that it had happened. I attribute that to UN forces being present in a lot of locations in Mogadishu: manning strong points, maintaining security, exercising force protection, conducting aggressive patrolling, and other such activities.

METHODOLOGY

- **Monitor daily**
- **Coordinate with joint/combined staffs**
- **Weekly site assessments**
- **Cover with NGOs & PVOs**
- **Must go to the commanders**
- **Staff must be responsive to needs**

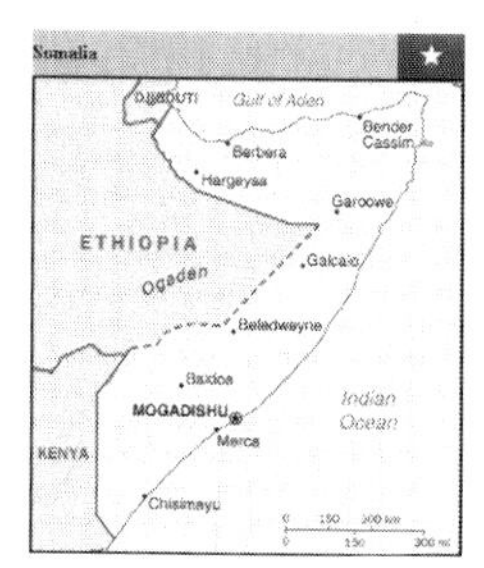

This was our methodology. We monitored the situation daily and I'll add that we were not only doing this in Mogadishu but throughout Somalia. Constant coordination with both the joint and combined staffs was necessary so that they were aware of what we were doing. We also had to address cultural sensitivities, not only of Mogadishu clan members, but also of other coalition forces. We picked sites and went out every week to ensure forces were maintaining good security. We even provided classes and briefed the NGOs and PVOs. We gave them formal presentations on how to maintain their own security and conducted security assessments for them. Finally, security reports have to go to the commanders and the staffs have to be responsive to unit needs.

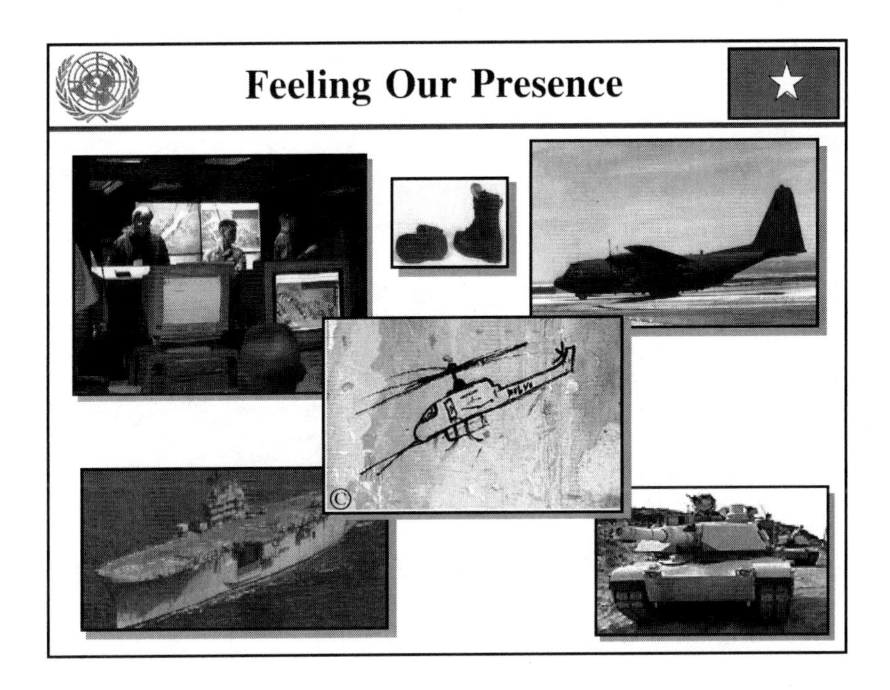

It was all about making our presence known. How did the Somalis know that we were there? We made ourselves visible. One force protection asset was ship platforms. The more things that you can put at sea the better; it provides this natural barrier around you called the ocean. With this barrier you are less likely to be a soft target for some of these asymmetric threats that are out there. How do you keep tabs on these threats? How do you monitor the environment? What new technologies are out there that enable us to see what's going on? Can we deploy video cameras in these environments? Where normally we might have a Marine with only his eyeballs, can we put a camera in that location and a force behind that camera and monitor? These were some of the things we looked at.

Look at the photograph in the top center—black boots. There is a story behind those boots. We landed in Mogadishu in December 1992 with the 15th Marine Expeditionary Unit (MEU). They had chocolate chips (desert camouflage uniforms), but they did not have the brown boots to go with that uniform. They had to wear black boots. They went in under Chapter 7. They had to protect themselves in several instances. The Marines did a professional job within the ROE. The word on the street over time was "Don't mess with the black boots." Do not mess with the black boots because they were Marines. Now let's fast forward to 1995. General Zinni is the commander of the combined task force United Shield. He's also the Commanding General, I MEF. He remembered from his experience as the C/J3 for UNITAF in 1992 what these black boots were all about. He chose not to issue desert boots because of the big psychological factor that carried over from three years before.

Another way of showing our presence was the AC-130. During United Shield we would fly the AC-130 at night, working in different areas and sectors. Aircraft had had such a profound effect on the Somali population in Mogadishu during 1993 that they painted caricatures of the AC-130 and Cobra. When the aircraft flew at night, the people couldn't see it, but they knew it was watching them and they knew it was backed by lethal power. The town got very quiet at those times. It was another way of showing our presence, another aspect of force protection. It wasn't always an individual Marine or soldier at a checkpoint.

Tanks. We saw the Grozny lessons learned on how to employ them in cities and how not to employ them. In this mission it was one of our best night systems. It was one of the few platforms that had thermal sights. We didn't have a lot of thermal sights. We now have them on our light armored vehicles, but did not at that time. Tanks are also armor-protected. The initial policy decision was not to employ them because it was thought that it looked bad when we were supposedly there for humanitarian reasons. Later on though, as we found out how complex the situation was and how great the security and force protection threats were, we did in fact employ them. We put them on mounted patrols at night to the point that we got a phone call one night from Aideed asking that we not run the tanks up and down his road as it was keeping his family awake. We continued using the tanks in that MOUT environment. A dismounted patrol in that environment can be mistaken for bandits. To this day we believe that one of our earliest casualties was such a case. One of our first Marines killed was on a dismounted patrol that may have been mistaken by someone on a neighborhood watch program. At night such patrols moved through an area very quietly. What happened was that a Somali protecting his home would see a shadowy figure and shoot. From that point on we used mounted patrols; they were very successful.

I am with the warfighting lab. What are we doing with technology enhancements to address problems identified in Mogadishu? The squad radio is one answer. We have heard others talk about how important it is to be able to communicate in urban environments. We're getting squad radios out to our forces right now and having success with the system. We'll get it down to squad level. You heard what the Chechens said about such radios: they would have liked to get even more of them. We found out through experimentation that it's very important for an infantry squad leader to be able to communicate with other squad members in his platoon in an urban environment. Other systems helped with situational awareness. One such device was called GOSSIP. You can use it to monitor those individuals who are bad guys by adding them to the database and distributing the information down to the lowest tactical elements. A sentry at his post can look at someone's ID and if he pops up on GOSSIP he realizes that maybe he has a war criminal or someone else we are looking for. Otherwise such an individual would just slip through our checkpoints. Other elements, urban camouflage and equipment. You see a photograph of a small, wrist end-user terminal. LtGen Knutsen, Commanding General, I MEF, is with us here today. His forces participated in Limited Objective Experiment 6 at Twenty-Nine Palms, California and tested such equipment. We successfully employed one of these during our experiment. It is one way we can get information down to squad leaders in an urban environment. We would like to connect it with a global positioning system (GPS) so we could provide other friendly units' locations. We all know that fratricide risk is especially high in an urban environment. You may not know you have friendlies on the other side of a wall as your force engages. What non-lethal systems can we bring to bear? The bottom center shows a portable reverse osmosis water purification unit. Part of force protection is dealing with the local population and helping them. The bottom right photograph is a multi-lingual translator. We've had some success with those in Kosovo. We would take them with us to assist in speaking with sentries from other nations.

Other things we are looking at: intelligent targeting systems. We experimented with these during Urban Warrior to help us locate and detail a target and then engage with precision weapons. VTAL UAV—in an urban environment you need to have something that can go up and down, in and out; that's what we are looking at there. We are testing micro-robotics, counter fire systems for snipers, and other robotic systems that can go in and out of buildings and check situations without putting a man in harm's way. We want Future Fighting Vehicle Systems that are smaller, more agile, and have better force protection than a tank. We need micro-UAVs that a small unit can launch to see what's on the other side of buildings. We can use night sights for snipers, individual thermal sights, and other such systems.

COMBAT DECISION RANGE

- 12 scenarios from peacekeeping to combat
- Computer video with unit facilitators
- Assess and train all leaders from squad to BN staff
- More decisions in one scenario than a week in the field
- 27% improvement in readiness

Something else that we've developed to help prepare for this is a combat decision range for squad leaders. The range consists of different scenarios spanning the spectrum from peacekeeping to combat. Here you see a CNN reporter putting a microphone in a squad leader's face. He's standing there in real time; he has to answer the reporter's questions. In another scenario the reporter has been taken hostage by the foe and the squad has to go on patrol and find her. The tasks include urban environments, medical evacuations, humanitarian assistance, intense combat, and other challenges.

PROJECT METROPOLIS

- Objectives:
  - Develop a comprehensive urban warfighting Program of Instruction (POI)
  - Develop TT&Ps to enable Marines to fight and win in MOUT with reduced casualties
  - Recommend improvements to existing and future MOUT training facilities
  - Evaluate selected enabling technologies that enhance small unit combat capabilities

Last, but not least, we are working on Project Metropolis. It's a follow-on to Urban Warrior the purpose of which is to address elements of urban operations like those mentioned by guest speakers discussing Grozny this morning: unit leadership and tactics, techniques, and procedures at the small unit level. How do we coordinate infantry and tanks with light-armored vehicles and AAVs in an urban environment during the "three-block war?" We're experimenting with that now. I MEF forces are supporting us again; we use Marines in operating forces so as to get an accurate assessment of where we are and where we need to go.

What's our future hold? This little lad is going to be carrying an RPG in eight to ten years. The thing we don't know is where he's going to be doing it. It's conferences like this that are going to help us provide mission success. We know we can't avoid these environments, though we'd love to. There's not a Marine here who wants to go into the urban environment and fight, but we don't make that decision. Others make it for us and I think all of us understand that very well. We're trying to address the inevitability that we're going to have to go in there. How do we do it? How do we do it better so we can accomplish the mission and bring everyone back? This is what we need to be addressing for today's and tomorrow's U.S. military.

Appendix F

THE URBAN AREA DURING STABILITY MISSIONS CASE STUDY: BOSNIA-HERZEGOVINA, Part 1

COL Greg Fontenot, U.S. Army (ret.)

1st BRIGADE
1st ARMORED DIVISION
TASK FORCE EAGLE

"READY FIRST COMBAT TEAM"
MOUT IN SASO
Bosnia—1995–1996

The intent of this presentation is to discuss the tactical, operational, and strategic goals implicit in Operation Joint Endeavor and to review the means that were used to attain these goals in a MOUT environment. The unit I commanded, the 1st Brigade, 1st Armored Division, led U.S. Forces into Bosnia and was assigned the responsibility for the strategically important Posovina Corridor, including the critical city of Brcko in northeast.

Overview

- Height-Weight Chart for the AOR
- Tactical, Operational, and Strategic Goals
- Center of Gravity
- Ways and Means
- Weight the Main Effort
- How was this different than stability and support operations (SASO) in the open countryside?

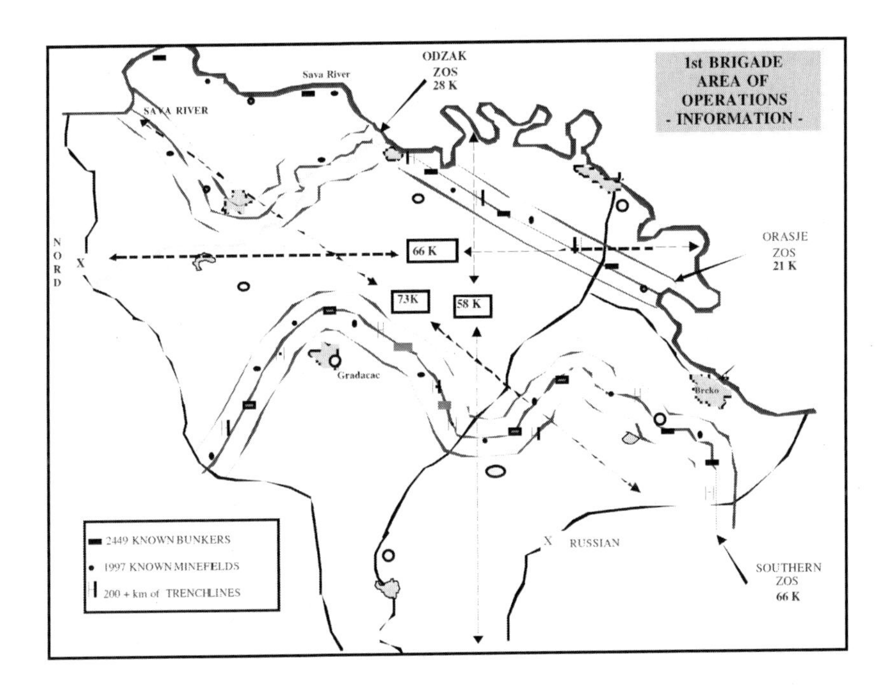

The brigade's area of operations encompassed 3,500 square kilometers and included citizens from the three major factions in the nation: Serb, Bosniac, and Croat. The region had been the site of long-term fighting, the remnants of which included those elements shown at the bottom left of this slide. Major cities and towns within this northern portion of Bosnia-Herzegovina included Gradacac (Bosniac); Brcko, Modrica, and Bosanski Samac (Serb); and Odzak and Orasje (Croat).

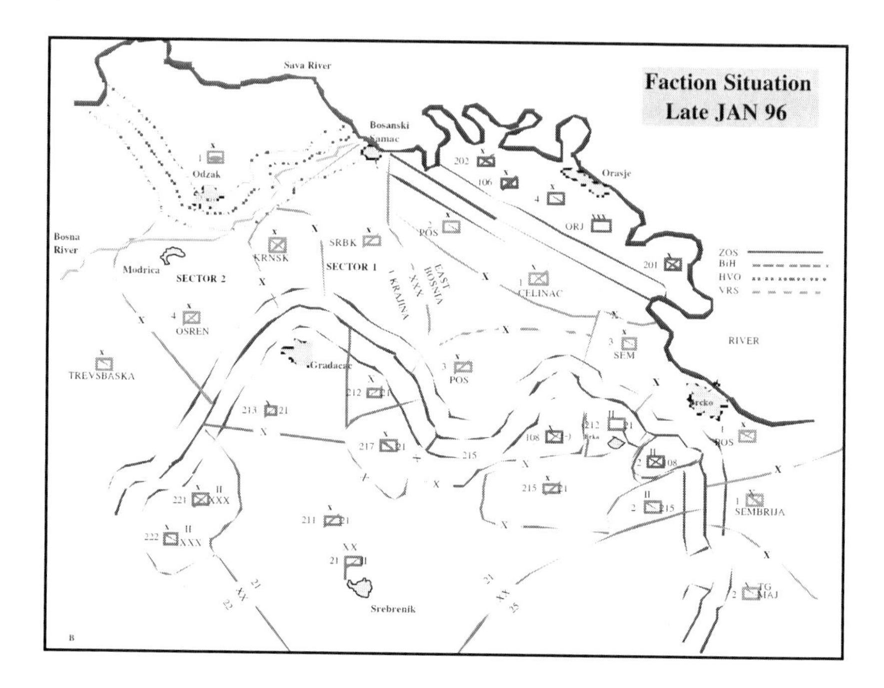

When the 1st Brigade arrived, its area of operations contained some thirty thousand armed troops representing the three factions, many of whom occupied static positions à la WWI.

Tactical, Operational, and Strategic Goals

- Tactical: Separate forces; gain control of the AOR—contact civil authorities and IO/NGO
- Operational: Disarm troops; make it hard to go to war (clear mines, blow bunkers, bury trenches); develop cooperation with civilian authorities and with IO/NGO
- Strategic: Routine compliance with GFAP

Remember—It's the economy stupid

Military tasks were well-defined in the Dayton Accords and well understood by both Implementation Force (IFOR) and the factions. While these tasks had a serious element of danger, they were reasonably easy to plan for, prepare for, and execute.

We quickly understood that because the civil support elements were slow to deploy it would be essential for our military force to assume critical civil tasks. We understood that, by necessity, the brigade would become involved in setting the conditions necessary for achievement of the civil aspects of the treaty as well as those military. The leadership did not view this as mission creep but understood that civil-military cooperation was an essential, implied task of the mission.

Successful execution of the military tasks set the conditions for peace by removing the immediate threat of actual combat, but only by achieving routine compliance, restoring the economy, and starting down the road to reconciliation could lasting peace be achieved.

The reality of the situation was that there was a compression of the three levels of war. Tactical actions, for example, could have immediate and significant strategic impact.

Center of Gravity

- If there was a center of gravity in the MND North sector, it was Brcko
- Large Serb displaced persons (DP) population, mostly from Sarajevo (31K post-war population was 95% Serb, but 30K Muslims wanted to return)
- Linchpin of the Posovina—links Pale to Banja Luka
- Bitterly contested during war and after Dayton

There were several real hot spots in Bosnia in 1995, to include Mostar, Bihac, Sarajevo, and the Posovina. Some observers argued that Brcko and the Posovina held the key to achieving the ends of the Dayton Accords. In the RFCT sector, Brcko was the key point on the ground. The city was the center of gravity both in military and economic terms; it was the place at which success had to be obtained.

Brcko: Center of Gravity

- All three factions involved
- Eleven known war crimes sites
- Symbolic for the blown bridge at the start of the Bosnian Civil War
- Ideal to defend—Sava River bounds the town on the north—one east-west highway—ZOS to the south—no easy bypasses or concealed approaches to the town

Assumptions

- MOUT is an environment and not a discrete mission
- A shot fired is a tactical defeat
- Use both reactive and proactive means to keep the lid on
- They were war weary at the out set, but looking for trouble by the end

At the outset the factions were sufficiently war-weary that they were accommodating, but that changed as time wore on. All sides began to foment confrontation by early spring 1996. Croats and Bosniacs focused their efforts on religious observations or graveyard visits in Sprska. The Serbs reacted to what they perceived as threats by the other factions with rent-a-crowds of howling Serbs that could be on the scene in 30 minutes.

The Ready First Combat Team (RFCT, the 1st Brigade, 1st Armored Division) lacked the force strength to effectively control any of the towns, let alone Brcko, by military means alone. While I believed that we had to have a presence in urban areas, and Brcko in particular, I sought to avoid becoming tied to fixed positions anywhere. Our approach was to adopt aggressive patrolling of selected checkpoints and to prepare for the isolation of towns if required.

Ways and Means

- Checkpoint at the bridge to close off east-west road; bridge had symbolic value as the place where the war started; was reasonably defensible; its control conveyed confidence
- Patrol regimen—7 days a week at random times, day and night
- Engagement with political and military leadership

Ways and Means

- Weight the main effort (3–5 CAV had smallest AOR, most resources, and priority of effort from brigade)
- When surprised or challenged, we sought to react more quickly than they could escalate. Remember that rent-a-crowd could be on the scene in less than an hour.
- Always remain firm and impartial to the point of aloofness

I weighted the main effort by assigning the largest available infantry force to Brcko and by assigning that unit the smallest area of responsibility (AOR). We also provided more resources to that unit than to supporting effort units. The brigade continuously worked both the civilian and military aspects of all relevant problems. I probably devoted 75 percent of my personal time and the brigade's effort to activities involving Brcko.

Troops Available

- 3–5 Cavalry was the mechanized task force at Brcko. It was equipped with Bradley IFVs and tanks, and augmented by military police.
- Civil affairs
- PSYOPS
- French parachute reconnaissance
- UK sound acquisition
- US covert

Troops Available

- JSTARRS GSM detachment
- Pioneer UAV GSM detachment
- Air cavalry
- AC-130 for point reconnaissance
- Fix-wing oblique photo reconnaiss ance
- Overhead imagery
- Tactical SIGINT and COMINT

Troops Available

- Field artillery—moved it around constantly—treated it as a presence force—artillery was the weapon of choice on all sides during the war—ours impressed them
- Target acquisition radar—could accurately detect small arms fire—then we followed up
- Dogs—mine dogs, bomb dogs, and patrol dogs

Military Means

- Proactive—continual development of intelligence via JCOs, CA, PSYOP, patrols and engagement, covert surveillance
- Reactive—security patrols, REMBASS, aerial reconnaissance (daily with internal means and frequently by other means), imagery, COMINT, overt surveillance, JMC, bilateral meetings, drive-bys, and home and away dinners

Aggressive patrolling and persistent efforts to maintain the intelligence estimate proved crucial to staying ahead of the factions. We remained engaged and vigilant both on the civilian and military fronts. The brigade built bypasses around Brcko so that we could assure access. We maintained a checkpoint in town that was both adjacent to the mayor's office and near the ramp to the Brcko Bridge. The intent was to be seen and to deny anyone the opportunity to destroy the bridge or threaten IFOR-IO traffic using the bridge.

Nuts and Bolts

I. TTP for reaction to any incident
   - Isolate
   - Dominate
   - Mass
   - Attack at all echelons

II. "Kobyashi Maru"—or as James T. Kirk said—cheat
   - Change the rules for high intensity technology
   - Don't set patterns

III. Predictive analysis = initiative

Engagement: Civil Authorities

- **Presence**
  Patrols: Day/Night, Mounted/Dismounted/Aircraft
  CA/PSYOP Drive-Bys and Assessments
- **Formal**
  Bilateral Meetings
  Civil Military Seminars
  Civil Military Projects
- **Informal**
  Dinners—Home and Away
  Office Calls
- **Opportunity**
  Periodic Press Conferences
  Special Radio Shows
  Joint Interviews

Measures of Effectiveness

- Some three hundred homes were started in Brcko suburbs between July and November
- Thirteen were blown up—in contrast to 65 in a single night in another sector
- Brcko Serbs with the most to lose stayed engaged throughout
- "Nothing happened" may be the desired result

The focus on Brcko had to be economic as much as military. To that end, the brigade organized and chaired the Posovina Working Group (PWG). The PWG brought all international organization (IO) and NGO representatives who worked in the corridor together to develop a vision for the end state in the Posovina and the means to achieve that end state. All IO agencies except the ICRC participated, as did nearly all NGOs. The vision focused on three main axes:

- compliance
- economic cooperation
- reconciliation

By the end of our tour, routine compliance was the order of the day. Some $44 million had been brought into the area and some hesitant but important first steps toward reconciliation had been achieved.

The Arizona Market was established at the zone of separation near the center of the brigade's area of responsibility. It became the model for the rest of Bosnia for stimulating economic activity and establishing a potential for economic reintegration.

Though thirteen homes were bombed, nearly three hundred were not. Brcko led the way in the return of DPs to their homes.

It is too early to tell whether these were the right steps, but they were often cited by locals, diplomats, and representatives from the IC as the most important initiatives in Bosnia.

Force Protection

- Force protection is a combat multiplier, not a mission
- Hunkering down in base camps is not only an ineffective way to advance the cause, but increases risks to the troops
- Maintain the standard for the duration—full battle rattle and alert troops was our standard

Contrasts Between Urban and Rural Efforts

- Intensity of effort higher in towns
- Frequency of patrols higher in town than out and continual presence maintained in Brcko
- Developed and planned in detail the means of isolating towns and moving forces from converging axis quickly
- More dismounted effort in towns than outside

Bottom Line

- METT-T works
- You have to think about the problem
- Do not need purpose built units—task organization works
- MOUT is an environment and not a mission

RFCT did a superb job in Bosnia at the relatively low cost of one killed and six wounded in action (WIA). We achieved what we did by considering MOUT as a condition of the mission and applied common sense and our doctrine to the problem. We maintained a continuous presence in Brcko, working toward military and economic solutions in cooperation with the various factional, IO, and NGO communities. While this kind of operation may be a corporals' war, it requires thoughtful and thorough leadership at all echelons.

Appendix G

THE URBAN AREA DURING STABILITY MISSIONS CASE STUDY: BOSNIA-HERZEGOVINA, Part 2

COL James K. Greer, U.S. Army

Lessons Learned

Fighting Crowds in the Cities of Bosnia
28 August 1997

On the 28th of August 1997, stabilization forces (SFOR) belonging to Task Force (TF) 1-77 Armor were conducting peace enforcement operations in the Serb Republic of Bosnia (RS) in conjunction with the International Police Task Force (IPTF) when they were attacked throughout their sector by large crowds of civilians. By the end of the day, the TF had defended itself against approximately 2,000 civilians and learned a great deal about stability operations in urban areas in the face of hostile crowds. The purpose of this brief is to pass on lessons painfully learned by some great soldiers.

The Asymmetrical Response

- Peacekeepers are:
  - Well prepared for armed resistance
  - Predisposed toward armed conflict
  - Constrained by ROE and social mores from shooting unarmed civilians
- Crowds armed only with sticks and stones are an asymmetrical response that confronts peacekeepers with challenges for which they are ill-prepared
- Crowds operate best within the confines of a city in which the terrain advantages accrue to the crowd vice the soldiers

Rent-A-Crowd

Clearly, the big tactical-level lessons learned from the 28th of August center on SFOR actions against hostile crowds. Many of these lessons learned are specific to how crowds operated in Bosnia. It is important to understand that these crowds were not spontaneous formations of people with complaints. They were, in fact, the result of a planned response to SFOR operations. Many of the people in the crowds, by their own admission, were paid up to 100 German Marks to demonstrate against or attack SFOR for the day, hence the name "Rent-A-Crowd."

Additionally, it was clear that these crowds were being used as an asymmetrical response to SFOR heavy mounted forces. RS leadership knew they could not hope to challenge SFOR conventional military operations. They knew the SF OR soldiers would not knowingly harm civilians or unarmed protesters. The Bosnian Serbs learned well the lessons of the Infatada in Israel and planned a deliberate operation against SFOR. What we expected was to spend the day observing police actions and possibly confiscating some illegal long-barreled weapons. What we ended up doing was defending against deliberate attacks directed specifically against SFOR and the IPTF.

How Crowds Fight

- **Weapons**
  - Rocks (often women & children)
  - Molotov cocktail
  - Sticks
  - Snipers
- **Tactics**
  - Attack in waves/cycles (Rocks, then rush)
  - Move through buildings
  - Block movement and reinforcements
  - Climb on vehicles

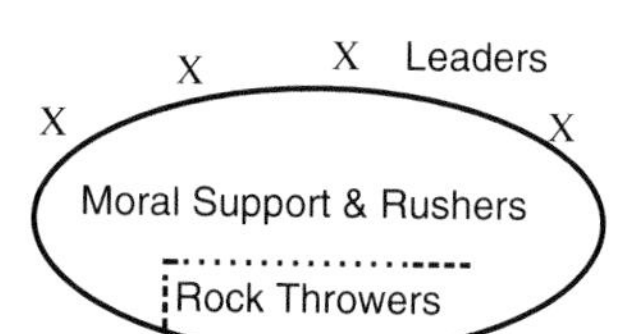

The crowds that attacked SFOR on 28 August did so without using small arms or any other military weapons. They were instead armed largely with sticks and stones. Later in the day, Molotov cocktails were used against our forces defending the Brcko Bridge in attempts to set our vehicles on fire. Crowds were composed of men and women of every age, including children and the elderly. Women and children were often used to throw barrages of rocks at our troops, based on the theory that we were unlikely to physically harm them. Military-aged men were usually the instigators or engaged in hand-to-hand fighting with our troops.

Crowds generally fought for short intense periods followed by lulls when they regrouped and rested. Usually the Serbs attempted to mass a crowd of several hundred against a single platoon of about twenty SFOR soldiers. Most periods of actual fighting were less than an hour at the end of which crowds would draw back to extreme rock range. They would sit down, rest, eat, and await the next set of instructions from the instigators. Occasionally, some Serb would throw a rock just for continued harassment.

Crowd Command and Control

- **Key Leaders** (Instigators)
  - Easily Identifiable (Red Armband or Flag)
  - Position on flanks and rear
  - Avoid direct confrontation
- **Communications**
  - Air raid sirens to assemble
  - Phone and local radio stations for instructions
  - Hand-held Motorolas for tactical control
  - Cell phones to political leadership
  - Music to adjust crowd efforts

For the Serbs, command and control of the crowds was extremely important and well executed. First, the crowds had to be assembled. This was accomplished by the simple expedient of turning on all the air raid sirens in the cities. This pre-arranged signal caused the crowds to assemble. The sirens were also the signal for the people to turn on their radios. The local radio stations passed instructions on where to assemble and what actions to take. Those people who didn't have a radio simply went to the local police station and were told what to do.

This effective process resulted in crowds of 500–800 people assembled and operating against SFOR within a half hour of the first siren sounding. For command and control at the crowd-level, several "instigators" were in each crowd acting as small unit leaders. They would pass the word verbally on what actions to take, whether throwing stones, rushing barbed wire, or backing off to rest and reorganize. Hand-held Motorola radios were used by the instigators and observation posts (OPs) for communications. Cellular phones were used by reconnaissance elements and the senior leaders not with the crowd. The final means of communication was music. Patriotic Serb songs were sung to raise the intensity of crowd demonstrations when desired by the Serb leadership. Taken together, these means provided excellent control of the Serb crowds.

Crowd Recon and Mobility

- **Recon**
  - Small cars for mobile patrols
  - OPs along routes
  - OPs overwatch fight from buildings
- **Mobility**
  - Know the terrain
  - Tactical on foot
  - Operational by bus
  - Build up forces over several days
  - Block opponent mobility (Dumpsters, trucks, debris, burn cars)

Deliberate reconnaissance was an integral part of the crowds' operations. Local police and civilians were set in OPs along the major routes that SFOR used. Because of our strict discipline in using only approved routes known to be clear of mines, after a year and a half where SFOR would go and not go was reasonably predictable. OPs could be simply a man in a telephone booth or a civilian couple sitting in their front yard. Civilians in small cars moved about the sector, reporting on SFOR movements and stopping to observe every time an SFOR column stopped. Within cities, OPs were set up with people just looking out of windows. All these actions combined to give the Serbs observation throughout the sector and excellent situational awareness, not only at the command level but at the crowd level also.

Tactics against our vehicles were well thought out. Men and women would lay down in front of vehicles, knowing that we would stop and not run over them. In towns, barricades were quickly built out of junk, dumpsters, timbers, or destroyed cars. These barricades channeled our vehicles into cul-de-sacs or directly into crowds. This allowed other elements in the crowd to attack with stones or by climbing onto the vehicles. Once on vehicles people would physically attack our soldiers or destroy or steal equipment (antennas, tools, or individual soldier equipment).

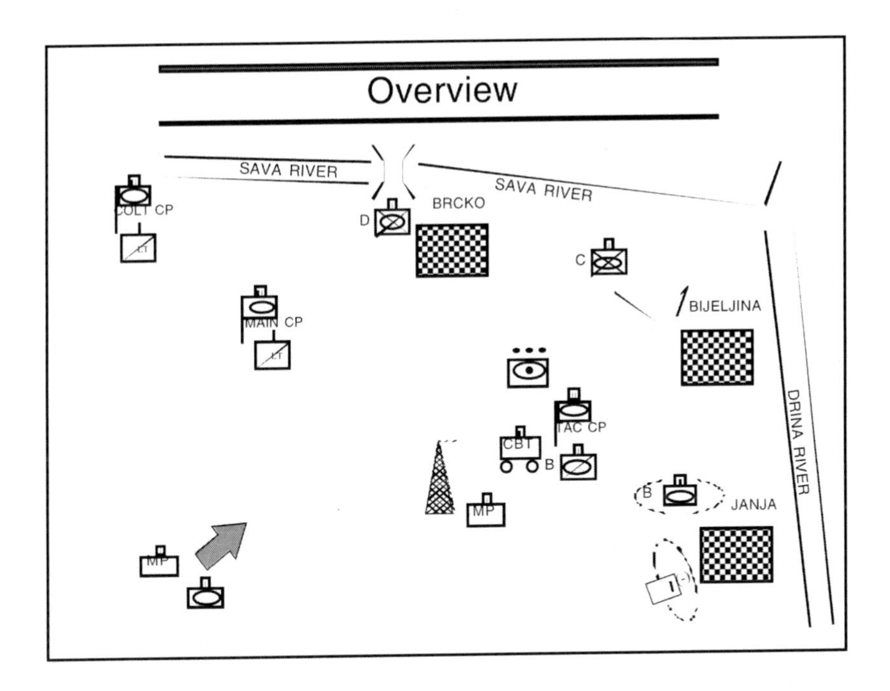

In August 1997, the Steel Tigers were operating in the critical region of the Posavina Corridor in northern Bosnia, including the key cities of Brcko and Bijeljina. Brcko is the most important city in Bosnia in terms of the implementation of the Dayton Peace Accords (DPA) for several reasons. First, Brcko is situated at a chokepoint where the Bosnian Serb entity of the RS is only four kilometers wide. Loss of the city to the Bosniac and Croat Federation would effectively split the RS into two pieces. Further, road, rail, and river lines of communication east-west through the RS and north-south connecting Bosnia with the rest of Europe go through the city.

The city of Bijeljina is important because it contains the headquarters of the Ministry of Police (MUP). In the RS, power is invested in the police more than any other government institution. Whoever controls the police dominates the RS. Additionally, the headquarters of the Special Police (anti-terrorist) Brigade of the RS, the most powerful armed force in the RS, was located just south of Bijeljina in the town of Janje.

SFOR forces were in temporary checkpoints (TCP), observation posts, and blocking positions at 0430 on the morning of 28 August when sirens sounded in the cities of Brcko, Bijeljina, and Janje. Almost immediately, crowds of approximately 500–800 civilians gathered in each city and began to attack SFOR throughout the sector. The remainder of the day was spent restoring order, extracting and protecting members of the international community (IC), and protecting our soldiers. The situation in each city was significantly different, producing distinct lessons learned from each event.

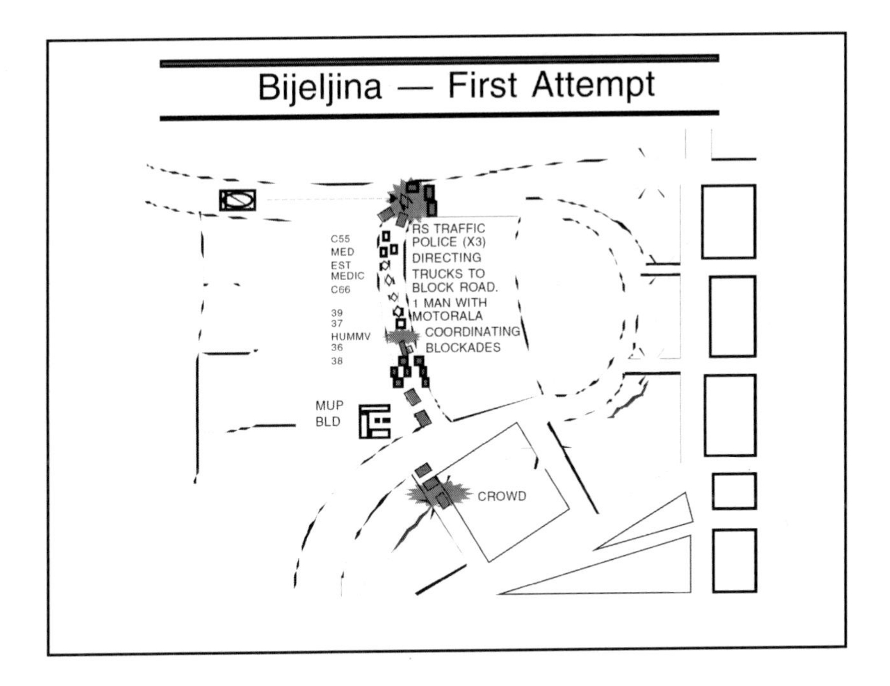

Operating in Bijeljina, C/2-2 Infantry had two tasks. First was the support of IPTF monitors as they attempted to enter the RS Ministry of Police building and conduct an inspection to locate illegal weapons. In order to support the IPTF, C Company had to move Bradley Fighting Vehicles (BFVs) and HMMWVs to overwatch positions cordoning the MUP. However, the entire MUP was surrounded by approximately 800 civilians armed with rocks, bottles, and sticks. Additionally, large trucks had been positioned in depth to block all roads. C Company made two mounted attempts to reach the MUP. Not only was the way forward blocked in both instances, but as soon as the SFOR vehicles penetrated the outer ring of crowds, trucks moved in from side streets and blocked the egress routes.

On several occasions BFVs were forced to cut through yards and parking lots to break contact. At other locations within the city, C Company was successful in their second task, which was to secure and guard three radio towers. Throughout the operation, SFOR Apache and Blackhawk helicopters provided critical observation down narrow streets. Although not the preferred solution, on one occasion helicopter rotor wash was used to assist the IPTF and TF Commander to break contact with a crowd.

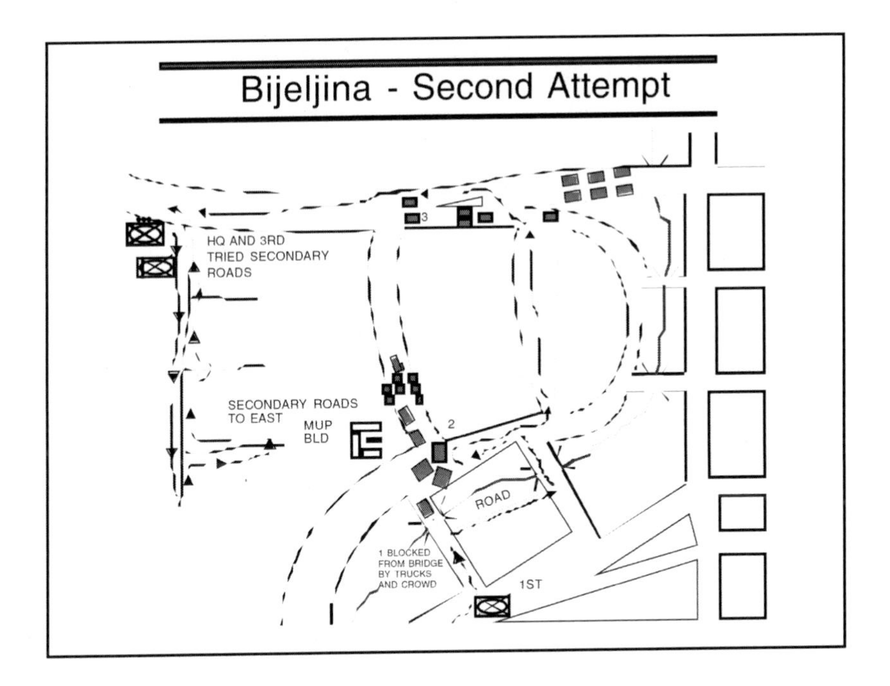
Bijeljina - Second Attempt
HQ AND 3RD
TRIED SECONDARY
ROADS
3
SECONDARY ROADS
TO EAST
MUP
BLD
2
ROAD
1 BLOCKED
FROM BRIDGE
BY TRUCKS
AND CROWD
1ST

Lessons Learned — Tm Steel

Improve
- Liaison with aviation to ID ro utes in and crowds
- Contingency planning

Sustain
- Flexible platoon sized movements
- Communications, both internal and external
- Reconnaissance at lowest level
- Information dissemination upon notification

Good Techniques	*Bad Techniques*
Multiple routes to disperse crowds	M2A2s on secondary road
Lead with armored veh to clear routes	Staging in open after 1st attempt
Continue to patrol throughout the event	

Several key lessons were learned from the operation in Bijeljina. First, the Bradley proved to be an effective vehicle for operations in built-up areas. The ability to pivot steer and traverse low walls, and its mobility enabled the BFV to extract itself from tight spaces. The height of the BFV let soldiers observe over the crowds and, coupled with the traversing turret, made it difficult for attacking civilians to climb up on the vehicle. In contrast, the HMMWV proved difficult to maneuver, requiring significant jockeying to turn around in the narrow streets and vulnerable to attack by people on foot. These vehicle observations were repeated throughout the day in Brcko and Janje.

SFOR helicopters proved invaluable for assisting the ground elements. Adequate coordination had been made at the TF level, but a key lesson learned is the value of face-to-face coordination at the maneuver unit (company in Bosnia) level. The ability of the helos to recon routes from the air and provide advance warning of crowd movements through cities is enhanced if the aircrews have a detailed knowledge of the company plan.

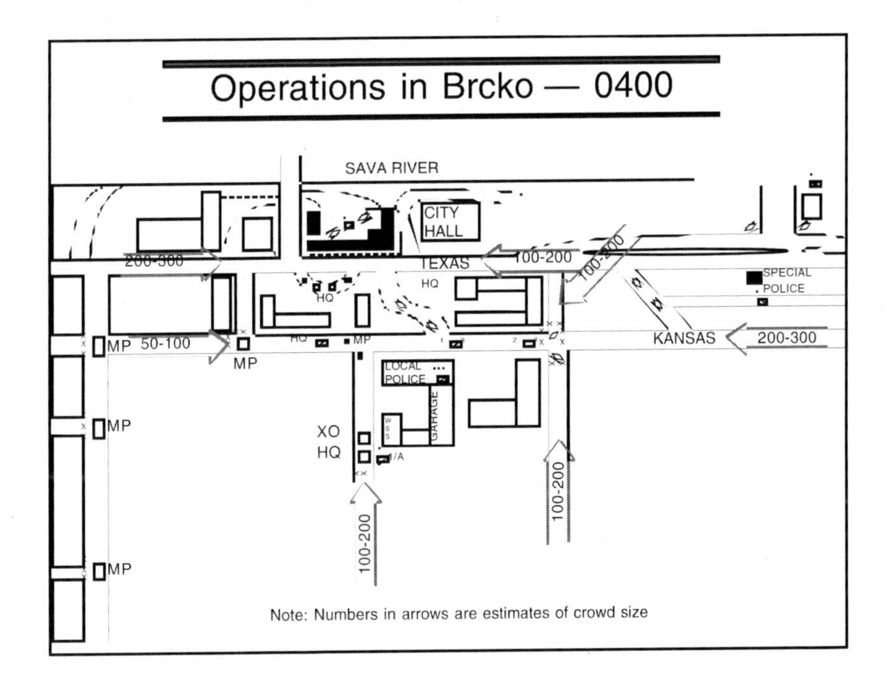

Team Dog initially deployed into observation posts and checkpoints in and around Brcko to prevent the movement of long-barreled weapons into the city and to support IPTF in the performance of UN-mandated inspections. As at Janje, around 0430 hours the city's air raid sirens sounded. Soon, angry crowds gathered and began to attack Team Dog throughout the city. As the various platoons were attacked and overrun by crowds of up to 800, they began to fall back toward the fixed position guarding the Brcko Bridge, one of only two SFOR lines of communication over the Sava River.

For the next twelve hours, the bridge defenses were surrounded and almost continually attacked by crowds wielding bricks, railroad ties, and eventually Molotov cocktails. The SFOR soldiers defending the bridge were subjected to almost non-stop bombardment with bricks and rocks, thrown in most cases by children and older women. Actual attacks on the bridge came with little warning and consisted of crowds attempting to penetrate the wire barrier and wall, climbing on BFVs and HMMWVs in attempts to damage them, and attacking the SFOR soldiers.

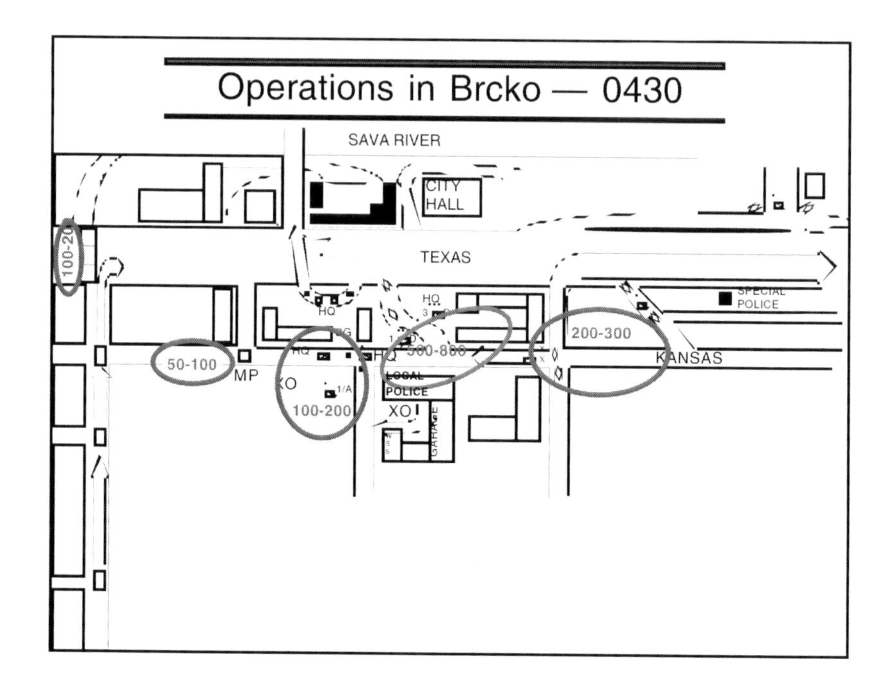

As at Janje, hand-thrown and M203-fired CS grenades proved too little to affect the crowd's actions. Throughout much of the day SFOR soldiers were in hand-to-hand fighting. An important lesson is the value of physical training and conditioning. Our soldiers were generally stronger than the Serbs and able to fight for a longer time without being tired. Key to this effort was CPT Kevin Hendricks' (Commander, D/2-2 IN) example and leadership. Realizing early on that the attacks on the bridge would last for an extended period, he periodically rotated his soldiers across the bridge to the Croatian side of the Sava. This had the twin effects of letting them rest, rehydrate, and eat, plus they were out of sight of the fight, enabling them to relax in a setting of lower tension.

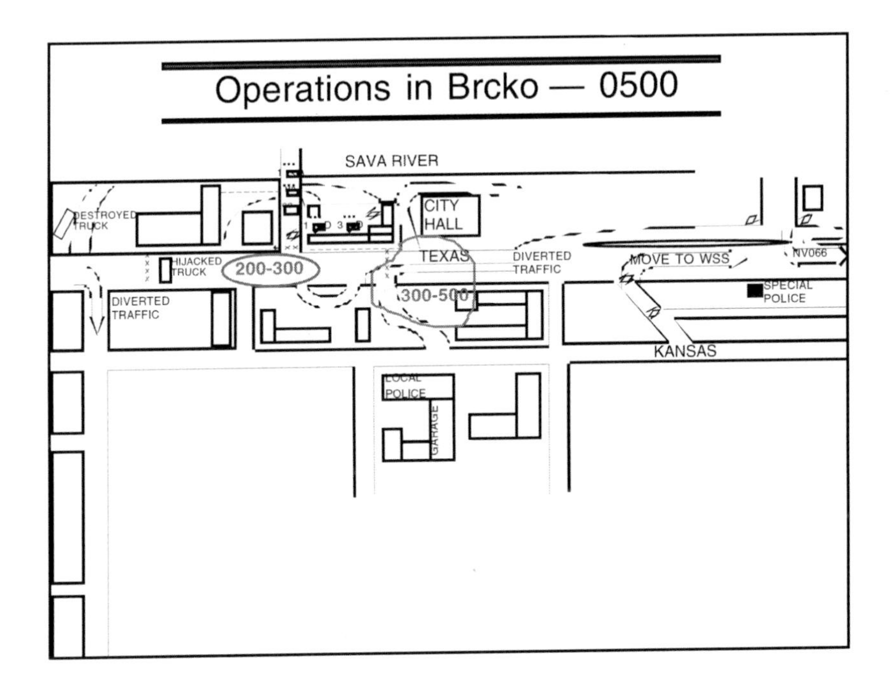

The crowd did not limit its attacks to the bridge. Several thousand Bosnian Serbs roamed the city of Brcko and attacked virtually any SFOR or UN personnel they found. First, they surrounded the JCO headquarters containing an eight-man team of American soldiers designated to be protected by SFOR. The JCO team was evacuated by a BFV platoon from Camp McGovern (located just south of Brcko). Key to success in the JCO evacuation was constant communications and use of a plan that had been pre-coordinated and rehearsed.

Besides SFOR, the IPTF was a key target in the Serb attacks. Over forty IPTF vehicles were destroyed and those IPTF policemen who could not escape were surrounded in the IPTF headquarters building. Late in the afternoon, 2LT White, commanding 2d Platoon of Team Dog, took his four BFVs into the city and rescued the IPTF. A lesson reinforced in that operation was knowledge of the terrain. Their patrols over six months had taught the platoon every road, corner, building, and alley in the city, enabling them to quickly get in and out. Additionally, the BFV was well suited to rescue with its compartment in the back providing a protected area for evacuation. The neutral steer capability of the BFV let the platoon get in and out of one lane streets quickly, out maneuvering crowds that were moving on foot.

Lessons Learned — Tm Dog

Improve
- Handheld Cannister CS Ineffective
- Stalled HMMWV & Backing Out
- POL Products in Bustle Racks and Sponson Boxes
- Concertina Wire Does Not Stop a Crowd
- Buffer Zone Around Perimeter Barrier
- Combat Camera Vulnerable to Attack

Sustain
- ID Location/Trigger of Warning Shots for COAX & 25mm
- Adjusted Weapon Level Posture to Fit Situation
- Emphasis on ROE and Graduated Response
- Retrograde Planned and Rehearsed
- Rotation of Individuals to Decompress
- Buddy System in Crowd

CPT Hendricks' use of the Rules of Engagement (ROE) and a graduated response was masterful. Shortly after the siege began, CPT Hendricks realized SFOR was probably going to have to employ warning shots to break up attacks. He determined that SFOR would fire pistols, then M16, then BFV coax and finally, if necessary, the 25mm chain gun in a graduated response to deter further action. Additionally, specific SFOR soldiers were designated to fire warning shots, given signals for when to fire, and told which targets they could shoot at with no harm to SFOR or the crowd.

As the day progressed, so did the level of response. Eventually in the early afternoon, Team Dog had fired 9mm and M16 warning shots but the crowd continued to attack and penetrated the wire in several places. Based on their rehearsed actions, the designated gunner fired a burst of coax into a nearby deserted building. That machine gun warning had a quick and sure effect on a crowd that from the war knew the power of machine guns. The crowd quickly withdrew and further attacks were limited to stones and Molotov cocktails. Discipline, training, planning, and rehearsals had paid off.

Defeating Crowds — What Works

- **Use mobility overmatch**
  - Bradley was vehicle of choice
  - Speed, mobility in urban terrain, height
- **Use mass** (platoon vs. 300 persons a nonstarter)
- **Confront from multiple directions**
  - Stress crowd C2; flexibility
  - Break up crowd mass
- **Absorb/diffuse crowd energy** (tire them out)
- **Barriers**
  - Double strand, tank-heated concertina
  - Reverse slope (rock and bullet protection)
- **Civil Affairs and PSYOPs in outer ring**
  - Defeat crowd at its source/base of power

From this challenging day some techniques were generated for dealing with hostile crowds. First, take advantage of the mobility mismatch. So long as SFOR can keep moving, the ability of crowds to accomplish their mission decreases. SFOR armored vehicles in particular can go cross-country or cross-town and force crowds to chase rather than attack. The second lesson is to employ mass. Twenty soldiers against three or four hundred won't work. Lots of soldiers and vehicles coupled with the next technique, approaching from multiple directions, will confuse and overwhelm even a large crowd. Crowds are not the type of organization that can rapidly react in multiple directions.

Next, a crowd has a certain amount of energy, both physical and moral. Crowds get tired if they expend their energy chasing vehicles or trying to overcome stubborn soldiers behind well-built barriers. The crowd loses its moral energy if it doesn't generate success early in the operation (100 Marks will only motivate someone so much). A tired crowd will just quit and either sit down or drift away.

Defeating Crowds — What Doesn't Work

- Helicopter wash
  - Can push a moving crowd/disperse stationary
  - Risky (buildings, wires, rocks)
- Negotiating with Rent-A-Crowd ineffectual
- Up-armored HMMWVs
  - Low, can't turn, limited mobility in urban terrain
- Tear gas; unless in large quantities (grenades useless)

Items that did not work so well included grenades, whether smoke, CS hand-held, or those fired from the M203. The weapons simply don't generate enough CS in an open area to make a difference. The backpack CS dispensers are far more effective and will stop a crowd in its tracks.

Negotiating with a Rent-A-Crowd is a waste of effort. SFOR negotiates well with citizens who have legitimate problems and when SFOR is seen as dealing honesty with those problems. Trying to negotiate when the crowd was hired and really doesn't have a legitimate problem, or isn't even from that area, will always fail. The "instigators" can be negotiated with on local, immediate issues, such as rescuing a hurt civilian, but they are not empowered to negotiate on the issues that are the root cause of the riots.

The last challenge is up-armored HMMWVs. As mentioned above, they don't maneuver well in towns, can easily be stopped and climbed on, and crews can't see out of them if the windows are scarred by rocks.

Information Operations

OPSEC
- Required to avoid pre-positioning of crowd
- Deployment to avoid early warning
- Need to bring in special staff (PAO/Chaplain) early

Command and Control
- Simultaneous, separate close fights
- TACSAT and helicopters vital for dispersed urban operations

Public Affairs
- Integration at multiple levels critical
- Consumes organizational energy
- Combat camera very effective

Electronic Warfare
- Real-time collection against Motorola and cell phone
- OSINT provides good information
- Need jam capability vs . FM radio and Motorola

Throughout the period leading up to, during, and after the 28th of August 1997, execution of information operations in support of SFOR remained critical and challenging. For example, OPSEC in Bosnia is exceedingly difficult. First, SFOR's mandate requires a visible presence. We are an overt vice covert force and deliberately so. Consequently, force protection through OPSEC is problematic. In this case, several techniques were employed. First, the planning group was kept small. This avoided possible leaks through interpreters, local hire workers, or other non-SFOR individuals operating in the sector. A key lesson learned during the planning effort: bring in selected special staff early, such as the Public Affairs Officer (PAO), Psychological Operations Officer, and Civil Affairs Officer, to ensure information operations are fully integrated. The risk is that the information operations efforts will damage OPSEC, but the payoff is being able to use the power of the media and civil-military interactions.

Another key lesson involved was the vulnerability of the Combat Camera Team. The Serbs were well aware of the threat of having their actions recorded by video and shown on TV. They attacked the SFOR camera crews as well as civilian news teams whenever possible. To protect the Combat Camera teams, SFOR learned to position them under cover but where they could continue to record, such as in a building or inside the back of a BFV or 5-ton truck.

Back-Up Slides

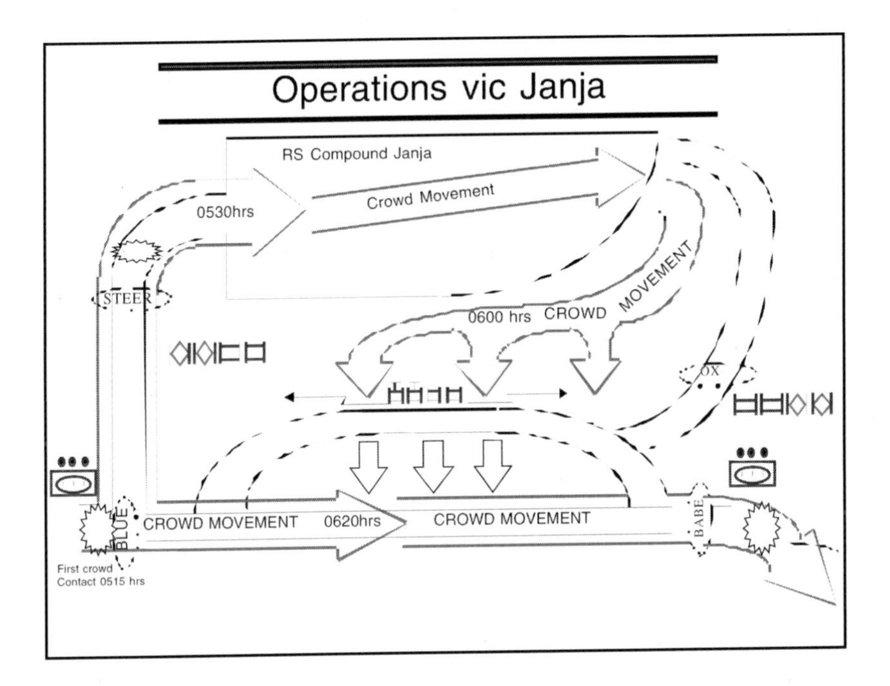

Lessons Learned — Tm Bull

Improve

- Communication and Radio Discipline
- Obstacle Emplacement
- Reconnaissance of AO
- Control of Aviation Assets in Support of Co/Tm

Sustain

- Commander's Parallel Planning with Staff
- Task Organizing Platoons (Ensure Tanks with Every Unit)
- Night Movement Allowed 30 Vehicles to Move with No Outside Distractions

Good Techniques	*Bad Techniques*
Tanks (Turbines & Gun Tubes)	Masking & Throwing Smoke
Double Strand Heated Concertina	Situational Obstacles
Mixing Platoons	Moving to Crowds to Disperse
Move Fuelers with Unit to Atk Pos	Pointing Weapons

Chronology

0300 hrs	RP JANJA
0330 hrs	ALL BPs ESTABLISHED
0400 hrs	PLT AT STEER ENCOUNTER RS POLICE - RECEIVES STATEMENT TO EXPECT 3000 - 4000 PEOPLE IN NEXT 2-3 HOURS
0510 hrs	AIR SIRENS SOUND IN TOWN OF JANJA
0515 hrs	CROWD MOVING TOWARDS BP BLUE (APPROX. 40-60 PERSONNEL)
0517 hrs	CROWD OVERRUNS BP BLUE - DISPLACE PLATOON TO BP BABE - BREACHER REPORTS "COUPLE HUNDRED PEOPLE ENROUTE"
0520 hrs	CROWD MOVES DOWN VILSECK AND SOME STAY AT BP BLUE
0530 hrs	PLATOON AT STEER MAKES CONTACT WITH CROWD - PLATOON IS SUCCESSFUL HOLDING 1ST WAVE OFF (APPROX 30-40)
0545 hrs	PLATOON COMPLETELY SURROUNDED BY CROWD (APPROX 300-500)
0550 hrs	PLT AT BP STEER THROWS SMOKE AND DONS MASKS - NO SUCCESS - SMOKE IS THROWN BACK AT TC IN VEHICLE
0600 hrs	CDR GRABS 3 OTHER TANKS TO FOLLOW HIM TO BP STEER
0605 hrs	CDR AND THREE TANKS ENCOUNTER CROWD AT BP BLUE - ATTEMPT TO BLOCK IN USING BUSES - START THROWING ROCKS
0608 hrs	PLT AT BP STEER GETS FREE AND MOVES ACROSS CORN FIELD
0610 hrs	COMPANY HAS ACCOUNTABILITY AND MOVES TOWARDS BP BABE
0615 hrs	PUSH ELEMENTS FORWARD TO BP BLUE TO RE-ESTABLISH BP BLUE
0620 hrs	CROWD MASSES AT BP BLUE FOLLOWS US IN TRUCKS, BUSES AND CARS
0700 hrs	CROWD BLOCKS OUR MOVEMENT ONTO ALABAMA USING SEMI'S AND CARS

LESSONS LEARNED

Improve

- Need More Documentation Capability (Video/Camera)
- Anticipate Asymmetrical Response During IPB
- OPSEC vs. Knowledge of Plan

Sustain

- Unit and Self-Discipline in Applying ROE
- TOC Operations
- Use of Civil Affairs/PSYOPs in Outer Ring
- Flexibility Based on Competent Platoons and Effective Company Command Elements
- Rapid Task Organization Capability
- Reinforcement From Out of Sector
- CONPLANs for Evacuatiing IOs

THE URBAN AREA DURING STABILITY MISSIONS: THE BRITISH EXPERIENCE IN KOSOVO

Brigadier Jonathan D.A. Bailey, British Army

MOUT
OPERATIONS IN KOSOVO
February – October 1999

Brigadier Jonathan Bailey
Chief Joint Implementation Commission
KFOR

RAND 22nd March 2000

The ACE Rapid Reaction Corps (ARRC) had been involved in Kosovo planning since 1998. However, this briefing will cover only the deployment period June–October 1999.

ISSUES PRIOR TO ENTERING KOSOVO

- EVOLVING MISSION
- NO ACTIVATION ORDER (ACTORD)
- FORCE GENERATION
- COMMAND STATES

The KFOR mission evolved through the following stages:

- A contingency for the extraction of monitors in a hostile environment
- Humanitarian relief
- Force protection in the Republic of Macedonia (FYROM)
- A variety of military options involving entry into Kosovo

What was actually carried out was, in effect, a relief in place to undertake a Peace Support Operation that was focused on the five major cities of the province.

The experience of the Kosovo force (KFOR) seems to indicate that in complex and rapidly developing political circumstances, the mission of a force is bound to evolve. Perhaps "mission creep" is what you call "mission evolution" when you cannot cope with it. If we controlled events we would not be in that position in the first place.

In a rapidly shifting scenario it was hard to generate an appropriate force. It would be made up of whatever nations wished to contribute. Without knowing what the precise mission would be, it was also difficult to be specific about what components were required, let alone deploy them in the desired sequence in a timely fashion.

Each nation committed its military force with a previously specified, closely defined command relationship with KFOR and a national support element.

KUMANOVO TALKS

- Complexity of interests involved and synchronization: NATO, National, UN

Photograph credit (this and all in this briefing): HQ ARRC/KFOR.

The problem was one of synchronization. The Federal Republic of Yugoslavia (FRY) would not agree to a deal unless it was subject to United Nations Security Council (UNSC) approval. A United Nations Security Council Resolution (UNSCR) did not exist and the North Atlantic Treaty Organization (NATO) was reluctant to have its deal with the FRY subject to a possible "watering down" at the United Nations (UN). Therefore, there needed to be a UNSCR in place before a deal could be agreed upon. The FRY wanted an immediate stop to the bombing and assurance that any NATO agreement would be subject to UN consideration and concurrence—not what NATO had in mind. Once stopped, could the bombing be restarted? NATO would have lost the initiative.

UNSCR 1244 was therefore produced while the bombing continued.

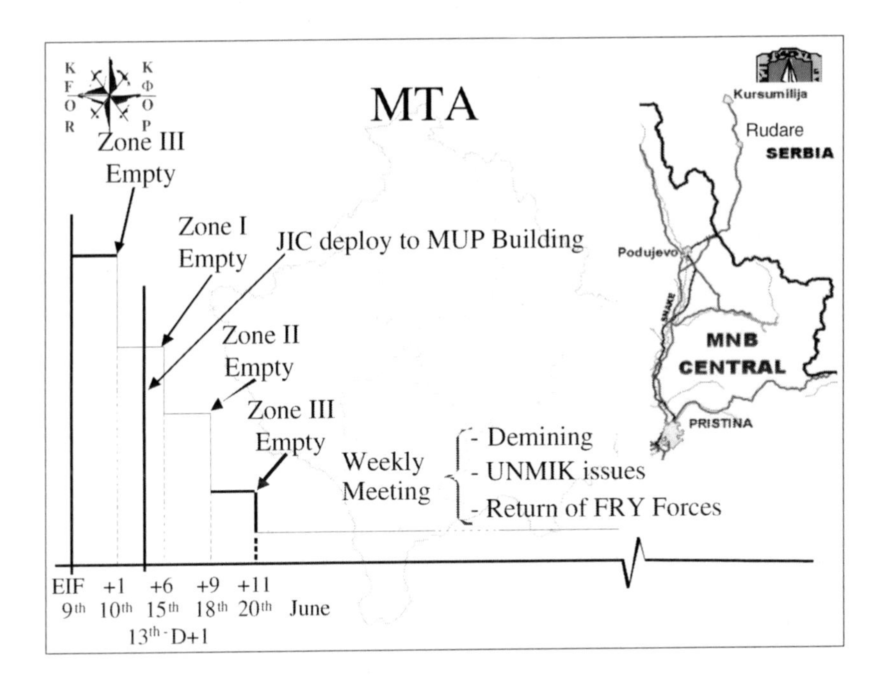

The Military Technical Agreement (MTA) required FRY forces to leave Kosovo by stages from mutually agreed upon zones. The zones were vacated in the sequence as designated in the MTA. Zone 3, in the north, was vacated immediately as a sign of good faith to secure the termination of bombing. Troops withdrawing from Zones 1 and 2 had to pass through Zone 3, thus it was also the last zone to be completely cleared. Once FRY forces had left Kosovo, KFOR held weekly meetings with them in Serbia to address continuing issues of concern.

D-Day designated the day that KFOR entered Kosovo.

WHAT WE HAD TO DO — THE STEADY STATE (UNSCR 1244)

- DETER RENEWED HOSTILITIES AND PREVENT THE RETURN OF FRY FORCES
- DEMILITARISE THE UÇK
- ESTABLISH A SECURE ENVIRONMENT
- ENSURE PUBLIC SAFETY AND ORDER
- SUPERVISE DEMINING
- SUPPORT UNMIK
- CONDUCT BORDER MONITORING
- ENSURE FREEDOM OF MOVEMENT

The operation to rebuild Kosovo had two preliminary elements. The first was to bring about removal of the FRY forces from Kosovo. That was achieved through the MTA. The second was the persuading of armed Albanians [the Kosovo Liberation Army/Ushtria Clirimtare e Kosoves (KLA/UCK); UCK is the designation of the KLA in Albanian] to disarm and disband. Once these two tasks had been accomplished, KFOR could get on with the task of trying to build a peaceful, prosperous, multicultural civil society.

While much of Kosovo is rural, it was clear that the most politically sensitive areas would be the main towns. Pristina was the center of provincial government. Possession of the city and the four main centres of local government (Prizren, Pec, Mitrovica and Gnjilane) was therefore symbolically essential. The issue was fundamentally about people; the major towns were quite simply where most people lived. Towns were the centers of communications and infrastructure. Apart from anything else, they were also the most accessible areas for the press.

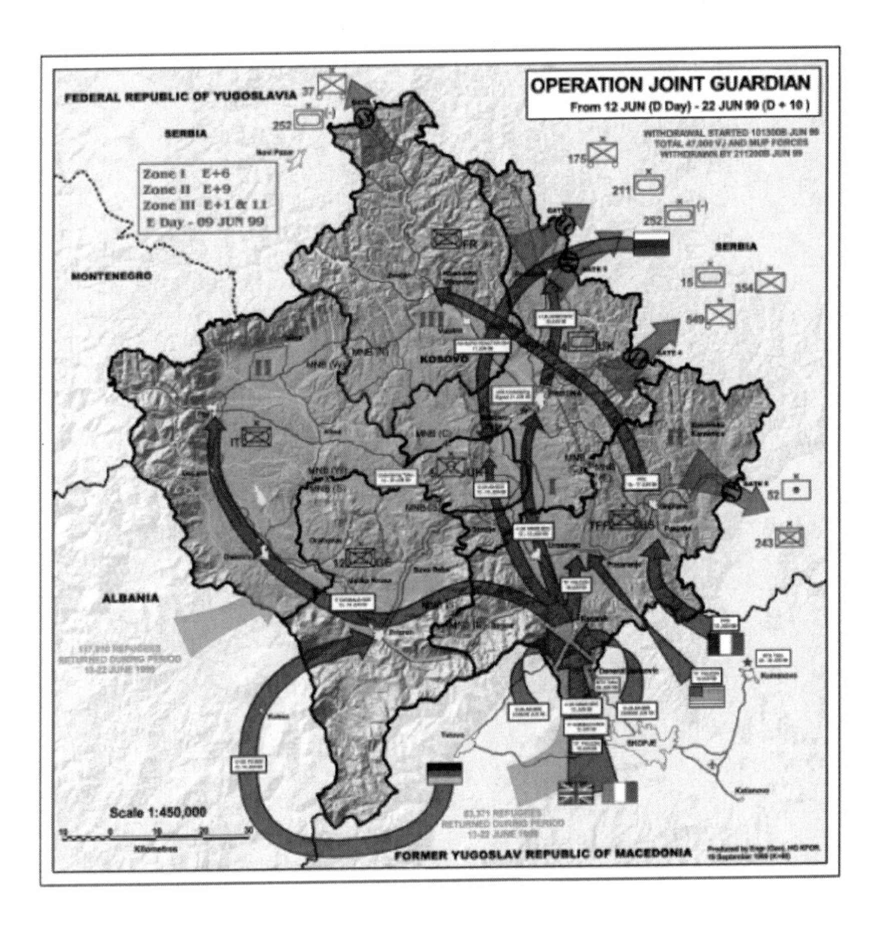

The KFOR main body entered Kosovo by the one good route: that from the south, the Kacanik defile. Thereafter its brigades fanned out to their areas of responsibility. This operation may have looked, and at times felt, like an advance to contact, but in fact it was more akin to a relief in place as KFOR's entry and the FRY Forces' withdrawal were closely synchronized to avoid a security vacuum (let alone a big traffic jam). The details of this complex maneuver were worked out between KFOR and FRY military staffs in the Europa 93 cafe at Blace over the two days following the signing of the MTA at Kumanovo.

VJ and MUP LIAISON

- The VJ and MUP representatives left Pristina on 20 June together with the remaining members of the FRY administration.
- FRY government representation continued throughout, providing the official point of contact in Kosovo .
- Weekly liaison meetings in Serbia started on 23 June.

The weekly liaison meetings were between the Commander Kosovo Force's (COMKFOR's) staff, senior FRY Foreign Ministry officials, and Yugoslav Army (VJ) and Yugoslavia Ministry of Interior (MUP) officers. The MTA required such liaison to be conducted by the Joint Implementation Commission (JIC); since the VJ and MUP officers were not allowed back into Kosovo, KFOR staff had to go to Serbia. Brigadier Bailey was COMKFOR's Chief JIC. He was therefore responsible for maintaining KFOR's ongoing relationship with FRY forces, as he was for relations with the KLA/UCK.

- Considerable concern that UÇK attacks on departing VJ would break the MTA.
- Negotiations started by AFOR in Albania in early June, leading to COMKFOR and Thaçi signing the Undertaking in FYROM on the night of 21 June.

The second of the two preliminary elements focused on the KLA. While the agreement between the KFOR and FRY was binding, that with the KLA/UCK was not in the sense that the MTA was. The KLA/UCK created a document that provided KFOR with a listing of measures linked to a timetable that the KLA/UCK undertook to carry out. This document was entitled the "Undertaking." KFOR acknowledged its receipt.

KFOR КФОР

UÇK UNDERTAKING

Stop all hostile acts

Close fighting positions

Move to AAs and establish SWSSs

• UÇK not of local origin withdraw

• All prohibited weapons and 30% of automatic in SWSSs

60% of automatic weapons in SWSSs

Demilitarisation complete

K DAY K+4 K+7 K+30 K+60 K+90

21st Jun

19th Sep

This KFOR-KLA/UCK Undertaking addressed the two crucial issues of demilitarization and transformation. This slide shows the measures as outlined in the Undertaking that constituted the sequence of demilitarization:

- Cessation of hostilities
- Assembly of the force in assembly areas (AA)
- The gradual handing-in of weapons at KFOR guarded secure weapon storage sites (SWSS) leading to the KLA/UCK's ceasing to exist on K+90.

The Commander, KFOR maintained regular liaison with General Agim Ceku, Chief of Staff of the KLA/UCK. Left to right: Brigadier Jonathan Bailey, Chief JIC; Lt Gen Sir Mike Jackson, COMKFOR; General Agim Ceku, Chief of Staff, KLA/UCK.

THE TRANSFORMATION OF THE UÇK

- The UÇK Legacy. Politics, Personnel, Weapons.
- The Kosovo Protection Corps
- The Kosovo Police Service
- Civilian Employment

The KLA/UCK Undertaking asked that the international community give due consideration to the transforming of the UCK into some other entity. Negotiations over what form that entity should take consumed the attention of KFOR for three months. Even after the transformation of the KLA/UCK there would be a legacy:

- The demilitarization of the UCK would leave its political wing in an ambiguous position.
- The UCK leaders would perhaps still be regarded as leaders of their community in some sense.
- Weapons would still be readily available to those who wanted them in that part of the world.

The Kosovo Protection Corps (KPC) was seen as a way of providing gainful employment for individuals in all communities who wished to help in the building of a new Kosovo. Its role is similar to that of civil defense forces in many other countries. It has no coercive nor law enforcement missions.

The newly created police force (not to be confused with the KPC) was the Kosovo Police Service (KPS). It was open to any individual, including former UCK personnel.

Many, especially students whose studies had been interrupted by hostilities, found scholarships around the world to help them return to the routine of civilian life. Others decided to go home and get on with making a living.

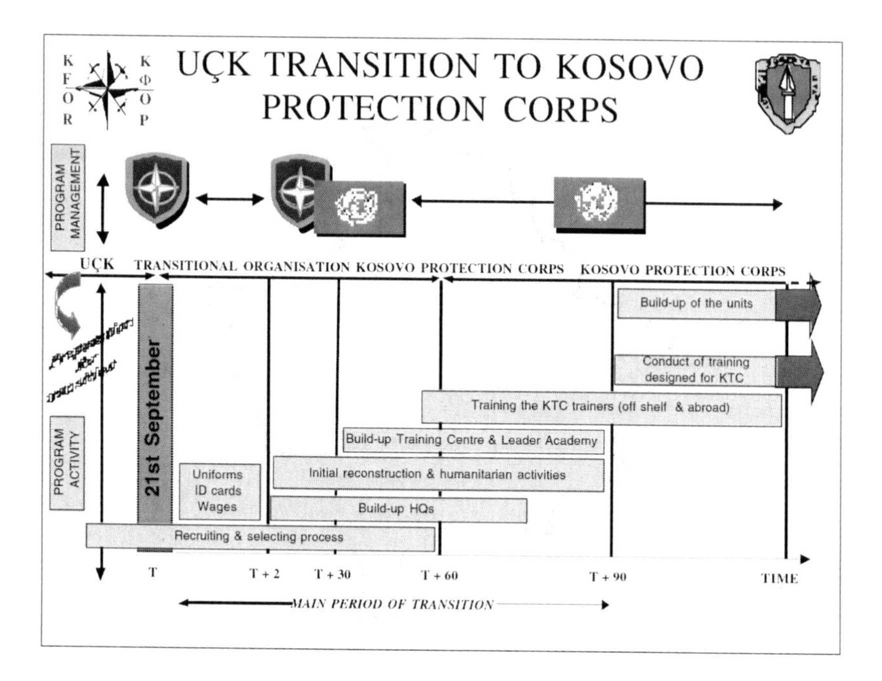

T-day was "transition day," the beginning of the 90-day process during which the KLA would transition to the KPC/KPS. By the end of this period there would exist a fully functional KPC/KPS, a change constituting the full transformation of the KLA/UCK as envisioned in the Undertaking.

THE FUTURE — WHAT TO WATCH

- TRANSFORMATION OF THE UÇK?
- UNITY OF EFFORT OF THE INTERNATIONAL PRESENCE?
- WHAT DOES FRY SOVEREIGNTY AMOUNT TO?
- RETURN OF FRY FORCES?
- MULTI-ETHNIC HARMONY?

Will the former KLA structure remain but go underground? Will it reactivate?

If the going gets tough, will international resolve (over 20 contributing nations) survive, conducting unified military operations with a common purpose?

Exactly how do we define FRY sovereignty and its bounds? What did UNSCR 1244 mean on the issue of sovereignty? What impression was given to the FRY in the MTA? How will the international community react to the outcome of an election in which every Albanian politician is likely to campaign for independence?

Do people in that part of the world really wish to live in multicultural societies?

MOUT ISSUES

- MISSION CREEP OR MISSION EVOLUTION?
- LEGAL ISSUES
- FORCE PROFILE

Mission Creep: When nations commit forces to an operation, they are in effect committing them to a variety of tasks which may be ill-defined at the outset. This is particularly sensitive when it comes to MOUT during peace operations in which the line between policing and military activity is seen differently by different nations.

Legal Issues: Soldiers are likely to operate in an environment of legal ambiguity and with different national interpretations. Are they involved in an armed conflict? What is their legal legitimacy? What are their powers? What is their personal legal liability?

Nations have different approaches to "force profile." These are held for particular national reasons and I will not make judgments. I will merely say that as a Brit, I am bound to favor the approach practiced by the British Army in Kosovo and elsewhere. The aim is to communicate with the people and only appear to threaten those who threaten you. The person in the street should see you as a friend rather than a frightening, anonymous soldier.

The following policies were implemented in support of this force profile policy:

- Do not wear helmets or body armor unless essential.
- Do not appear only when an incident occurs. Mingle on the streets.
- Dominate an area with foot patrols, especially at night.
- Integrate at low level. Live in peoples' houses with them. In the long run, this may be the best form of force protection measure.

At the same time, act decisively; be a credible military force, shooting to kill when appropriate.

In the early days, KFOR received an enthusiastic welcome from almost all of the population.

MOUT ISSUES

- MISSION CREEP OR MISSION EVOLUTION?
- LEGAL ISSUES
- FORCE PROFILE
- SOLDIERS AS POLITICIANS
- SOLDIERS AS THE CIVIL AUTHORITY
- SPECTRUM OF THREATS
- CONTROVERSIAL TASKS. Unlimited Liability.

Soldiers as the civil authority: KFOR was responsible for security while the UN Mission in Kosovo (UNMIK) was the legitimate government of Kosovo. However, until UNMIK was able to assume that responsibility, it fell to KFOR soldiers to assume the burden. It is not acceptable for soldiers to say that it is not their task when urgent, immediate action is required to uphold the overall mission. Soldiers must decide who owns buildings, houses, and factories. Who owns the fridge that people are arguing over and for which someone has just been shot? If trucks appear at a metal factory and start loading millions of dollars worth of ingots, are the "suspects" securing their property or stealing someone else's?

The soldier has to take a position because someone could be about to get shot over the incident. Who is going to run the infrastructure: mines, power, water, telephones, rail, hospitals, banks, TV, radio, newspapers, schools, and universities? Such issues as power and water supply failures cannot be deferred. On KFOR's first day in Pristina and other towns, people were being killed in disputes over these issues. These are as much G3 as G5 issues.

This sort of problem became mainstream business for KFOR and it is likely to be so for any similar force unless there is a civil administration waiting, formed up, and ready to deploy on the same timeline as the military.

Spectrum of threats: Troops must handle situations involving mines, snipers, riots, barricades, and celebratory gunfire. Are they dealing with political violence or criminal violence? It can be hard to tell the difference.

A particularly controversial task (soldiers and law enforcement): What exactly is the dividing line between military security and civil security, between soldiers' work and that of the police? If there are no police, then is it all a soldier's job? It is no good saying that as soldiers we don't carry out these tasks. People will die in front of your eyes if you do not and the mission could resultantly be compromised.

A force commander faces real problems if troops assigned to him for a peacekeeping mission come with restrictive rules of engagement (ROE) regarding, for example, the authority to take part in crowd control. It only takes one second for crowd control to become riot control. Are all troops deployed on peace operations necessarily highly trained for countering civil disturbance? Even if they are, is their government prepared to allow them to take part in politically sensitive operations involving crowd/riot control? There is much common doctrine on war fighting. Do NATO nations have a common doctrine on riot control and a common cultural understanding of the concept of minimum force and the graduated application of force against civilian crowds? If you are commanding a force of NATO troops and, say, Russians or troops from the United Arab Emirates, can you be sure that your orders will be interpreted and applied in a consistent manner in a complicated and sensitive political and legal situation? As the commander you will be responsible for their actions.

Crowd control . . .

can lead to riot control . . .

or worse.

MOUT ISSUES

- MISSION CREEP OR MISSION EVOLUTION?
- LEGAL ISSUES
- FORCE PROFILE
- SOLDIERS AS POLITICIANS
- SOLDIERS AS THE CIVIL AUTHORITY
- SPECTRUM OF THREATS
- CONTROVERSIAL TASKS. Unlimited Liability.

Guard Duties: A large number of KFOR troops were necessarily tied up guarding Serb patrimonial sites. These "tactical" activities actually assumed an "operational" significance.

Community Protection: After force protection, the primary concern was combating the intimidation of minority communities, especially Serbs and Roma. Typical methods of physical intimidation included murder, kidnap, arson, defenestration (throwing people out of windows), and drowning. Other methods were used to threaten communities by denying them medical support, education, blowing up electricity lines to certain towns (such as Strpce), and refusing to serve minorities in the shops and markets of Pristina. Sometimes a telephone call proved to be as powerful as a gun.

Should a peace operations force encourage or discourage the creation of cantons/ghettos? Are troops, deployed around these, protecting the occupants or holding them prisoner? Should humanitarian convoys supplying these be protected or not? If the occupants want to leave, should they be escorted? If they are, has the force become an agent of ethnic cleansing? If they are not and they are killed, has the force become an agent of ethnic cleansing in another sense? Troops can counter this in reactive mode or try to preempt it. For example, in Pristina many troops moved into apartments that housed threatened residents.

APPLYING THE LESSONS LEARNED: TAKE 1 (PROJECT METROPOLIS)

COL Gary Anderson
U.S. Marine Corps, Marine Corps Warfighting Lab

A Road Map for Urban Joint Experimentation

Our Advice to the Joint Community

- Start with an experimental concept. It doesn't have to be perfect.
- Experiment sooner rather than later. Nike has it right: "Just Do It."
- Get a real world JTF nucleus involved.

We have suggested to the Joint Advance Warfighting Program that we become partners in moving on urban experimentation and that we do it quickly. We don't need to spend days and days and weeks and weeks and years and years in building an experimental organization to go out and do this stuff. It's not that hard. It's just a case of rolling up your sleeves and going to work. We have a number of concepts out there. The Joint Working Group has some good work on paper that probably needs to be tested too. We put together our thoughts on some advanced MOUT concepts. The Air Force has given some thought to precision bombing in the urban environment. I call it 30 seconds over Mogadishu, but my Air Force buddies don't like that. The bottom line is: we need to get a real-world JTF involved so that we can look at the kinds of decisionmaking that would have to be done during actual contingencies and get on with the process of urban experimentation. What we're really into now is problem solving. We largely know what the problems are at each level of war in the urban environment. Now we need to get out there and solve them.

A Road Map for Joint Experimentation

- Step One: Use an existing venue; Yodaville is loaded into JCATS.
- Build a complex scenario to experiment against; don't make it easy. Include all three blocks of the 3-block war.
- Incorporate lessons learned from
  - Urban Warrior
  - Project Metropolis
  - MOUT ACTD
  - Joint Non-Lethal Weapons Directorate

There are some venues for experimentation out there. When we built Yodaville to try to increase Marine proficiency in urban operations, we put Yodaville in JCATS. That means you can maneuver forces in JCATS while also conducting live operations over Yodaville. And there are many lessons learned from the MOUT ACTD, Project Metropolis, Urban Warrior, and some of the work that the non-lethal weapons directorate has done. The bottom line is, there are facilities out there for use in experimentation. George Air Force Base is a pristine facility. They had just finished its upgrade when they made the decision to close the site. It is out there and available for use. We'll be conducting a run of Project Metropolis there in May 2000. We will have a lot of experience in using such facilities for experimentation; the offer of partnership is there.

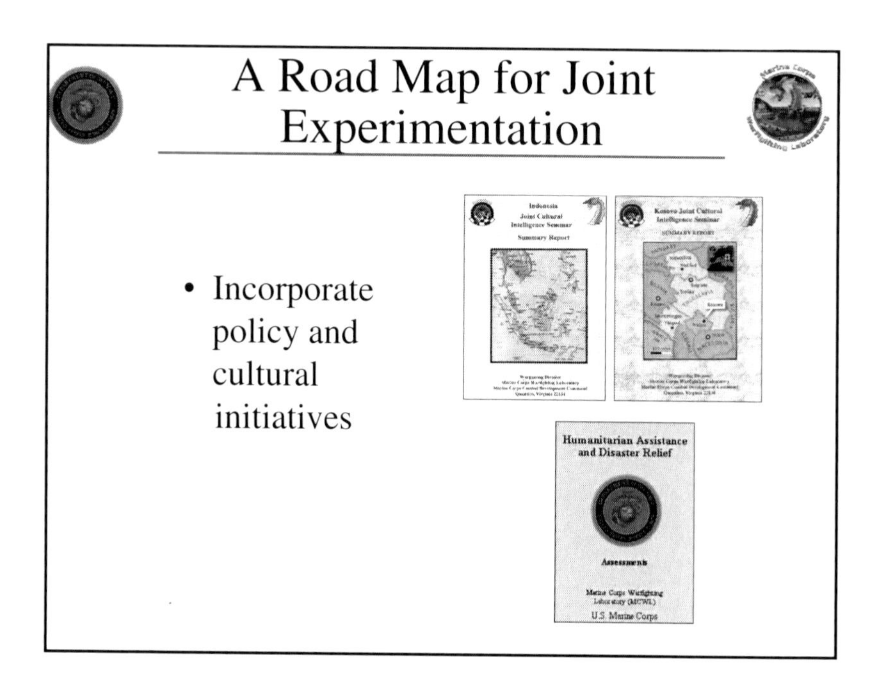

We have been working on cultural intelligence seminars. People mentioned that today. Knowing what kind of environment you're going into and how people are going to react to you are very important. We've had some good work done with humanitarian and disaster relief missions, which often take place in built-up areas. We have the opportunity to get out there and move into the next generation of problem solving. I would suggest that what we want to do is to attack this problem at three levels.

Pillars of Joint Urban Experimentation

Tactical
Mini- ACTDs as new equipment comes online

Operational
Distributed experimentation at actual locations (Yodaville and George AFB) with JCATS integration

Strategy – Policy
Seminar War Games

We really have some problems at the strategy and policy level. Although we advertise that 80 percent of our post–Cold War Marine Corps operations have been in built-up areas, only about 15 percent of those have actually involved anything approaching full-scale combat. What that means is that we've done a pretty good job of working problems out. We've got Ambassador Oakley here to talk to that piece tomorrow. We need to get better as an interagency group at doing assessments, figuring out what the end state may be, and trying to figure out how to properly bring one of these operations other than war to a conclusion. That's going to be a big part of policy and strategy. A lot of that can be worked out in good, solid, well-designed war games. You then create the structure and the sandbox in which to do your distributed experiments. Finally, you get your chance to solve the problems on the ground with the operating forces.

We still have a lot of things that need to be fixed at the operational level of war. In the Marine Corps we don't have any inherent, regular psychological operations or civil affairs capabilities. Civil affairs is in the reserve. We depend largely on the Army for PSYOP.

That we need more in the way of such capabilities is a "lesson learned" that comes out of every operation other than war, but we haven't done anything about it. We need to solve that. Another area of concern is nonlethal weapons policy. We're going to have directed energy weapons. We may not ever be allowed to use them under current policies. We carried very, very capable lasers, dazzling, not blinding weapons, that we could have used against snipers in Somalia. We were not allowed to use them for policy and legal reasons that still escape me. Those problems have not yet been solved. If we don't come to grips with them we will be in a position where we have some very capable weapons that we're not allowed to use.

Those kinds of issues remain to be dealt with at the JTF level. Putting this whole joint package together and using the best things that each of the services can bring to the table in a synergistic way is something that still needs a lot of work.

Finally, at the tactical level there are a lot of things that we would call frustrated requirements as we approach the MOUT ACTD wrap-up in September. We know what we need, but we don't see the right tool out there yet. What we need to do is have a continuing series of mini-ACTDs as people come along with potential solutions to frustrated requirements. We would then have a structure for taking a look at them and evaluating their utility so that we can start to crack some of these tactical and technical problems. For example, we still do a crummy job at tactical night operations. I've heard a couple of people mention it here in the room. That remains the case. Although we're getting a lot better during daylight operations, we've got a lot of work to do with urban operations at night. We heard a story about mock Stinger missiles [from Art Speyer]. We saw the same thing in Somalia during the hunt for Aideed. Aideed started a rumor that he had Stinger missiles. We had a squadron of Russian helicopters contracted to fly for us. They were immediately grounded. Not because the pilots didn't want to fly; they did, but since they were a contract organization, the insurance rates shot through the roof. We count on a lot of that in UN operations today. Quite frankly, Mohammed Aideed was probably the first individual in military history to achieve air superiority by correspondence.

We've got to figure out a way to preclude that. Making our helicopters survivable is another challenge. The helicopter is a wonderful weapon, but it doesn't do you any good if somebody knocks it down with a rock or an RPG. There's a lot of work to be done and we need to get at it. A number of people in Washington are asking for joint experimentation and if we don't get our collective acts together and figure out how to do this, somebody's going to tell us how to do it. I think it behooves all of us to move quickly.

A primary finding of our urban experimentation and operations: the strategic corporal is a very important concept. When you get into these types of operations, one kid with a gun doing the wrong thing can do an awful lot to harm to the operation if he's not properly trained. I remember very distinctly one Thursday morning in Mogadishu when we were confronted by a rent-a-crowd, exactly what COL Greer was talking about. Enemy snipers would hide in and amongst the women and children. This Thursday's was a particularly ugly demonstration; they brought out ladders that looked like those the Mexicans used in the movie *The Alamo*. They pressed up against the Turkish forces on the wall of the embassy and Capt Campbell, now Maj Campbell, and his FAST Platoon were on top of the embassy.

We were standing there talking about what we were going to do and both Capt Campbell and I were putting clips in our pistols, preparing for the close combat I hoped would not come since I hadn't qualified on my weapon for a couple of years. The embassy security officer rushed up to us and said, "They're up to something." Probably the understatement of the day. He said, as he looked at the ambassador, "If they come over the wall, the Marines are going to have to try to arrest them." And the ambassador looked at him and said "If they come over the wall, we're going to stack them up like cordwood." And we were. The bottom line is, we were really having to depend on good, well-trained, well-disciplined troops that knew what they were doing and were prepared to do it. Major Campbell will tell you tomorrow a little bit about a situation in which we had to take our snipers and execute what was essentially a strategic action at the tactical level. I believe we're going to see a lot more of that in the future.

That's a very quick tour of where we are in urban experimentation. I think the watchword from here on in is going to be "problem-solving," figuring out how to do this better and more efficiently, but also figuring out how not to get ourselves in a shooting situation when we don't have to. That's my story; I'm sticking to it.

Appendix J

TRAINING FOR URBAN OPERATIONS

MG David L. Grange, U.S. Army (ret.)

TRAINING & READINESS FOR URBAN OPERATIONS

Dave Grange
U.S. Army (Retired)

Today I would like to discuss preparation for combat and stability and support missions in urban terrain. I will relate what we did in Bosnia and then how we conducted reconstitution and combat training after redeployment to Germany while at the same time supporting or rotating units in and out of Macedonia. Most of my experience had been with light forces, light infantry, and ranger or SF-type counter-terrorist units until my three years with the 3rd and 1st Infantry Divisions during which I was a heavy guy. As we go through this, I will try to pass on lessons learned, good and bad.

Importance of MOUT

- We have always operated in urban terrain
- We still are
- We always will
- Whether we like it or not
- We have no choice but to be ready

There is no doubt regarding the importance of being prepared to operate in urban terrain. I believe that we have not put the training and equipment emphasis on urban operations that we should. There are periods during which units focus extensively on MOUT, but there is far from the level of training we actually need. I put a lot of focus on MOUT as a commander at the company, battalion, regimental, and divisional levels, but it was not enough. I did not do as much as I should have; I know I could have done better.

Civil War

"The Rebels could only be dislodged by desperate fighting among the buildings of the town. It was impossible and impracticable to attempt to relieve the press by throwing troops into the streets, where they could only be shot down, unable to return the fire. The Rebels fired on us from windows and doors and from behind houses. We had no choice but to rush the houses, breaking in doors to drive them from the town. My regiments did not falter, but in future I should much prefer to fight in open fields, than amid buildings and town streets."

Norman J. Hall, Colonel
17th Michigan Volunteers
Fredericksburg, Virginia
11 December 1862

Urban fighting has been with the United States Army for a long time.

World War II

The 2nd Battalion, 26th Infantry was given the mission to clear the center of Aachen. The commander organized his battalion into three hard-hitting company task forces.

- Each rifle company was reinforced with three tanks or tank destroyers
- Artillery was fired parallel to front of U.S. assault line; fire was brought in very close to friendly positions
- Machine guns commanded streets along axis of advance
- Assault troops moved through buildings using mouse holes
- Tanks fired on building ahead of each advance
- The battle took eight days with 80% destruction to city

We studied the World War II battle of Aachen extensively when I commanded the 1st Infantry Division. We conducted staff rides, walking battlefields with our commanders and sergeant majors to derive lessons learned about what we were going to do if we had to face combat in built-up areas. I want you to notice the task organization for the battalion that conducted the attack on the center of Aachen. Note the armored vehicle mix, the use of artillery and machine guns, the movement of troops through buildings rather than the streets. I believe some of the combat veterans from Somalia would tell you that during the Task Force Ranger fight only five percent of the casualties came during close quarter combat in buildings; most of the casualties came during movement in the streets. Also take notice of the 80 percent destruction of Aachen, which kept friendly casualties down. Aachen was a very set piece, deliberate fight conducted over a period of eight days. Contrast that battle with our MOUT experiences today and our extensive rules of engagement.

Operational Reality

- Environment
- Terrain
- Operational Spectrum
- Heavy vs. Light Units
- Equipment
- Threat
- ROE

The operational reality today is that rules of engagement will not allow us to conduct operations in a manner that causes extensive collateral damage or large numbers of noncombatant deaths.

I love snipers, but when getting ready for operations in Kosovo, the division had zero snipers authorized for the organization. I had sniper weapons in the arms rooms, but the table of organization and equipment (TO&E) had no personnel assigned to match men to the weapons. Note that I am only using snipers as an example. There are a lot of equipment issues, a lot of training issues that need to be addressed if we are to be prepared for urban operations. Snipers was one particularly dear to my heart because I believe in the power of snipers.

The reality for our division in Germany was that we had operated in and would continue to operate in urban terrain. We operated in Bosnia, in Macedonia, and we knew that we were either going to do a peace support operation or go into combat in Kosovo. Our primary war focus area was the Middle East, but our immediate concern was the Balkans. There the terrain was restricted; it included mountains and forests as well as numerous villages and cities. As a heavy unit, we had both advantages and disadvantages when operating in this type of terrain. We had the protection and mobility that our combat vehicles afforded (though mobility was quite often a challenge due to infrastructure weight restrictions). What we really lacked in sufficient quantity in our heavy division was foot infantry, not like my previous experiences in light, airborne, and ranger units where I had plenty of infantry to do the tasks necessary. We had too few infantry to control an area during peace support operations and too few to clear buildings and patrol if we were sent into combat in an urban environment. We lacked much of the equipment light units habitually have to support operations in urban terrain. We knew our adversaries would capitalize on their asymmetric tactics, techniques, and procedures to negate our strength in built-up environments. Finally, it is important to note that ROE, tactics, and approaches employed by fellow coalition members often differed from our own.

In Bosnia, my Task Force Eagle command was located in the north of the country. Colonel Fontenot and Colonel Greer have both discussed this geographical area in detail. I will therefore focus on a few points they brought up that hit home for me. Bijeljina was really a gateway into Serbia. The city of Brcko was crucial; it linked east and west Republic of Serbska and Zvornik [down the Drina River on the east side of our area of operations (AO)] bordered on Serbia as well. A lot of movement came and went through the LOCs in both Brcko and Zvornik, which were also choke points. And, finally, a fourth choke point straddling LOCs was on the west side of our AO at a place called Doboj. All four of these cities were key terrain for us. This meant that we were going to have to work in built-up areas. Lines of communication (LOCs) were restricted in and around these cities. This challenged our forces during movements and resupply through these points. The problems were complicated in that we were constantly dealing with a number of different demographic groups. The Republic of Srbska had an ongoing insurgency between opposing factions while we were there. We were caught between these factions while at the same time having to deal with disagreements between Muslim, Croat, and Serb groups. The environment was full of twisted relationships due to a subversive network of political leaders, black marketeers, police, special police, paramilitary elements, individuals with local agendas and, of course, the indigenous military in which different brigades took different sides.

Bosnia, cont.

- Threats
  - IW
  - Crowds
  - Attacks
  - Possible threats
- Intelligence
- IO
- Training voids
- Troops available
- Equipment issues
- Non-lethal
- C^3
- Proactive vs. Reactive
- Combat multipliers

A terrorist model for organization was used by the paramilitary groups. It consisted of a cellular structure that we knew we had to penetrate. The belligerents had reliable commercial communications down to the lowest levels; in fact, in some cases their ability to communicate was better than our own. In addition, many adversaries employed disinformation, deception, psychological warfare, and electronic warfare. They exploited civilians. They did all of this to level the playing field with our military, our international governmental organizations, and the non-governmental organizations with which we worked. Crowd use became a tactic of choice. Factions employed rent-a-crowds consisting of coerced mothers and children and criminals let out of jail for a day. These people were paid 90 Deutschmarks to conduct a riot. Paramilitary and special police were the instigators and controllers of these mobs. There were periodic bombings directed against military and international organization vehicles and buildings, usually in retaliation for war criminal snatches that periodically took place in the different sectors. There were verbal and written threats against members of the Office of the High Commissioner, the UN High Commissioner for Refugees, and the international police task force. Our intelligence system had to become more sophisticated; we had to employ police-type analysis, peeling back the layers of the urban environments to uncover information. We focused on breaking the links between money, corrupt leaders, paramilitary elements and the police. Our JCOs (special forces elements) were invaluable in this environment. We brought in people like Colonel Tim Heineman who is here today. We already had a joint military commission. We had a pretty good handle on the conventional side. What we lacked was an understanding of the paramilitary capability, so we brought in people with the experience like Tim. We made him the Joint Paramilitary Commissioner and used him to develop measures of effectiveness. We needed to know how well we were doing with our indirect, asymmetric-type operations. At times it was not readily apparent whether or not we were attaining success. You need someone that is focused on that area alone to make sure that you are appropriately influencing an area of operations in support of your mission.

Information operations (IO), especially civil affairs, PSYOP, and public affairs, was a combat multiplier for both the factions and coalition forces. Increased operational security and the use of deceptive measures served all sides well. We were denied the use of nonlethal means initially. It took several emergencies to obtain those means, such as the emergencies in the August 1997 Brcko riots. In fact, approval for riot control gas use during the Brcko came late, and thus the delivery means that we had was not optimum for that environment. In later discussions with people who worked during the L.A. riots, we found it had been easier for them to get tear gas release authority in Los Angeles than it was in Bosnia. We did have some nonlethal means for the communications tower take-downs. The towers in the Republic of Srbska were controlled by Pale, the shadow government. They would not allow the elected government to get on the air. Our mission was to provide equal time for both factions to use the media. We therefore had to seize the towers, but we did not want to use lethal force. The solution? Ax handles. There was a significant psychological effect when the crowd saw soldiers with their weapons slung and ax handles in their hands. The mobs had been using bricks, stones, and clubs. It's funny how arming soldiers with ax handles leveled the playing field psychologically. The crowd did not advance. We positioned soldiers with 40mm nonlethal rounds on the crowd's flanks in anticipation of having to use that capability to accomplish our mission.

Nonlethal weapons definitely were a deterrent. General Zinni, a marine and a hero of mine, states that properly employed nonlethal weapons make a force more formidable, not less so. We also know that they are perceived to be more humane, that they provide a means to exercise more control over a situation, provide flexibility to the commander on the ground, and allow application of a graduated response.

We had several organizational shortfalls that we had to correct after we got back from Bosnia. We had an insufficient dismounted ground reconnaissance capability. Vehicle reconnaissance was easily monitored by an enemy and was restricted to roadways due to unmarked minefields. We had to get reconnaissance on the ground. We had to get surveillance in position. We had to get reconnaissance in old buildings to watch intersections. We had to do all this in a clandestine manner. We didn't have the equipment to do that. We used the Norwegians' mine dogs, which were excellent at picking a safe path for soldiers after they rolled out of a vehicle and moved into a building. We would use engineer overt mine clearing support as a cover to get people in buildings where they could watch and report. We started seeing some success with such methods but it took time to measure the effectiveness of the effort. Due to these experiences, we concentrated more on foot reconnaissance after our return to Germany.

As mentioned, we had no sniper capability. We developed a sniper detachment in every battalion. We still had significant problems. For example, the TO&E only authorized day scopes for sniper rifles. Things like that had to be corrected immediately. We received a lot of support from special forces organizations and Ft. Benning to fill equipment shortages and to assist with training. We school-trained our snipers and developed a sustainment training program. These men not only conducted sniper and counter-sniper operations, but were also trained to observe and report in support of intelligence collection activities.

I talked about the need for a nonlethal weapons capability such as riot control gear. We finally received the equipment and issued it to the units. Therefore we did not have to wait for it to be brought down from a base in Hungary while a situation was getting out of hand.

We had to refocus our intelligence procedures and reorganize the G2 shop in order to deal with this kind of environment. We wanted to seize the initiative rather than react, so we stressed predictive analysis. We went from an order of battle type analysis, determining the type of vehicles and how many there were in a containment area, to a police-type intelligence system with supporting analysis.

Our goal was to get adversaries to dance to our tune. We conducted air assault operations using false landing zones (LZs) to make the enemy believe we were putting people into an area and cause them to "light up" their communications capabilities. Our electronic warfare (EW) assets would then target these communications. We conducted road watches. Something as simple as putting someone in a hole on the side of a road reporting license plate numbers to the G2 who analyzed who owned the cars helped us better develop our intelligence picture. New movement patterns for counter-reconnaissance were developed. We looked at innovative operational approaches; we tried to dislocate the adversary's activities. For example, the Sava River bridge in Brcko became a magnet for demonstrations. We therefore ceased 24-hour manning of the bridge so that it wouldn't act as a magnet for media and mobs looking for visibility. We went to patrols and manning the bridge at unannounced times. The creation of mistrust among our adversaries supported our mission accomplishment. When we seized a warehouse full of contraband, a building that was known to be controlled by the local police chief, we used photographs of the materials in an effort to develop rifts between the police and supporting military units. Long-range photographs taken by Apache helicopters showing people taking weapons out of their car trunks and concealing them behind their backs during a bus delay operation were shown to the media, thereby discrediting claims of peaceful intent. We worked the media hard. Our soldiers established a radio station and used it in conjunction with PSYOP and civil affairs. And we did our best to address the conditions underlying many of the people's complaints. One of our most powerful tools in attempting to favorably influence the population in the Republic of Srbska was a puppet show put on by our soldiers in local schools. It had an effect as powerful as a tank platoon, an example of how all power is relative and situationally dependent.

YOU'LL GO TO WA R THE WAY YOU
ARE TODAY...

NOT THE WAY YOU WANT TO BE

It's a challenge to get organized, to get equipped, to get trained and psychologically prepared for combat or peace support operations in an urban environment. Regardless of your personnel shortages, the time that you have available to train, or the resources that you have on hand, you've got to get on with it.

THE SOLDIER
IS SIMULTANEOUSLY
THE STRONGEST AND WEAKEST
PART OF ANY SYSTEM

Technology is important, but it's not the focus. The human dimension is where the training focus must be. Technology can enhance the preparedness of well-trained, tough, motivated soldiers. The soldier, not the technology, is the fundamental element.

SOLDIERS SHOULD NOT DO SOMETHING FOR THE FIRST TIME IN COMBAT

We did not know when we would have to fight in urban areas, but we knew that we would operate in a MOUT environment. We wanted the learning to be done in Germany, not in Kosovo.

Kosovo

- Ready for what?
  - PSO
  - Combat – restrictive terrain
- Will have to operate in villages and cities whether PSO or combat
- Limited 3-dimensional capability
- Heavy division constraints
- Lessons learned
  - Intelligence
  - IO
  - NL
  - Response
    - Sniper
    - Mortar
    - Crowds
- Preparation

In the guidance given for operations in Kosovo, the 1st Infantry Division was told to be ready for peace support operations or combat. We had to be ready for both, so we trained for both. Kosovo has restricted terrain and tough defiles difficult for a heavy unit to get through. Lines of communication would have to go through built-up areas and mountainous terrain. We had a limited air assault capability which meant that after insertion troops were limited to foot mobility (aside from some light wheeled vehicles). Our heavy division had just converted to Limited Conversion Design XXI (LCDXXI) which meant its tank and Bradley strength had been reduced. On the plus side, we gained more foot infantry and reconnaissance assets. Unfortunately, we also reduced the number of mortars in each battalion, a concern as mortars are the indirect weapon of choice for urban fighting. Our greatest concern was enemy mines, snipers, RPGs, and mortars. I really did not worry much about enemy tanks and APCs. We took lessons learned from Hue in Vietnam, from Beirut, Somalia, Aachen, Panama, Bosnia, and other places. We learned from British actions in Northern Ireland and worked with the British and Germans to put our training programs together. Most of our focus was on preparing junior leaders, lieutenants and squad leaders, because they had received little urban operations training before arriving in the division.

Leaders — Deadly Effective with the Elements of Combat Power

- Maneuver
- Firepower
- Protection
- Leadership
- Info operations
  - In required operational environments throughout the spectrum of operations

We focused on the elements of combat power. Here I have added information operations as a fifth element. I believe that's true not only in stability and support operations, but also in combat. This is the case whether our force confronts symmetric or asymmetrical threats.

In preparing leaders to deal simultaneously with stability, support, and combat missions, we had to teach them how to think about and adapt to uncertain environments; for example to change from peace support to combat in seconds. Even though a force is deployed on a peace support operation, it still has a combat requirement. And, they had to understand the second and third order effects of their actions. Good judgment was critical down to even the most junior level.

Training

- Germany
  - CMTC/GTA
  - Hammelburg
- Reconnaissance
  - AOR
  - Need for
- Ammunition
  - Training
  - For real
- Equipment

We had excellent MOUT training sites in Germany. They were not as advanced as the Joint Readiness Training Center (JRTC), but they were good sites. At the Combat Maneuver Training Center (CMTC, in Hohenfels, Germany) there was a complex battlefield similar to what we expected to find in Kosovo. CMTC had at least seven villages we could use for such training. We also conducted live-fire exercises with armor-mechanized company teams in the Grafenwehr training area. The German infantry school in Hammelburg was another invaluable asset in our training program. We conducted leader reconnaissance from sites on the borders with Serbia. We had to find additional ammunition for the training we conducted, like close quarter combat training using instructors from the Special Forces and the NATO School. Equipment shortfalls had to be remedied. Ft. Benning, Georgia helped us in standardizing signal devices used for building clearing. We developed MOUT kits with mechanical and demolition breaching capabilities. We had to get ladders. We had no knee and elbow pads. We fortunately had a limited amount of ranger body armor. I think the initiatives undertaken to acquire the Interceptor Body Armor and similar protection will go a long way. It amazes me that with our aversion to casualties we have not fixed the body armor issue. It's critical during MOUT.

Training, cont.

- Support
  - Ft. Benning
  - NATO LRRP School
  - Special Forces
  - USAREUR
- Combined Arms
  - BOS
  - Specialties
  - Task Organization

As I said, the infantry school at Ft. Benning, Georgia helped us in establishing our sniper school. We transitioned our Bradleys to what are called Operation Desert Storm Bradleys in which all soldiers in our three-squads per platoon MTOE [Modified Table of Organization and Equipment] could sit. We worked all of the battlefield operating systems into our training. And we worked a lot with civil affairs. We also mixed tanks, Bradleys, and other vehicles with foot infantry and engineers during training.

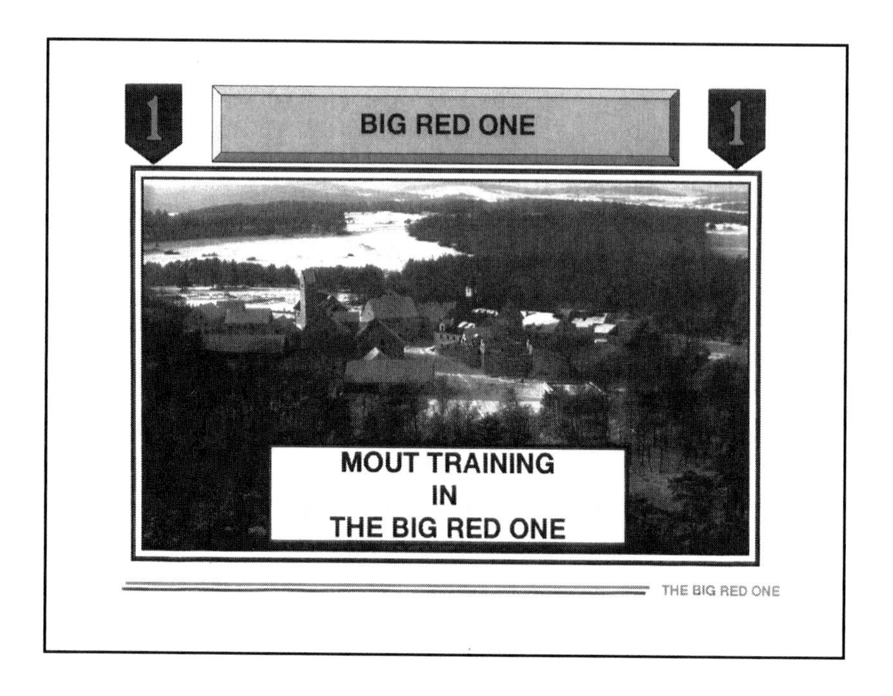

The entire chain of command was committed to this program.

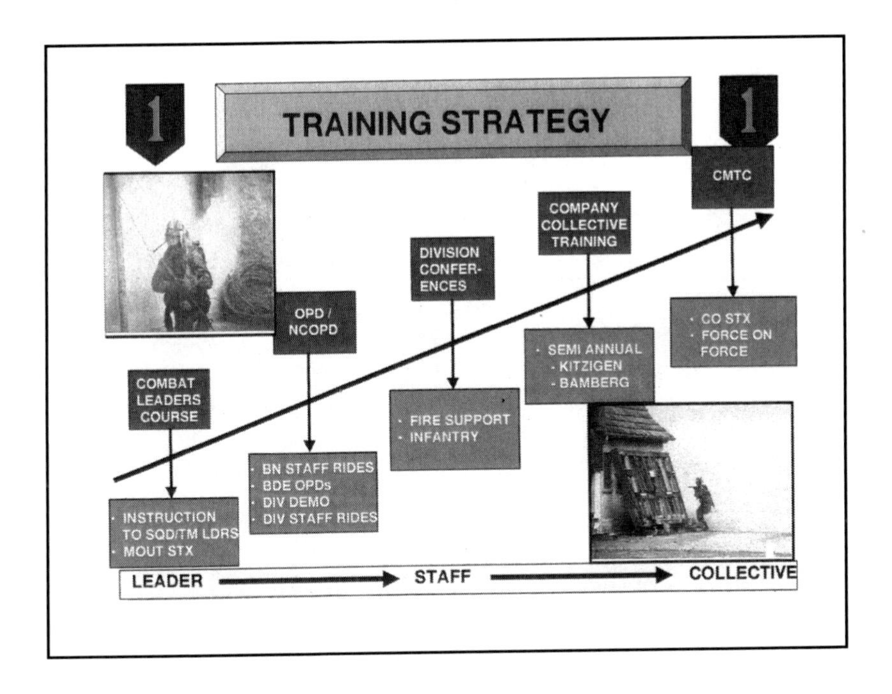

Leaders were trained first. NCOs and lieutenants completed a combat leaders course that we developed with help from Ft. Benning and the NATO Long Range Reconnaissance Patrol (LRRP) School. We also conducted officer and NCO professional development courses. The division's warfighting conferences focused on our new LCDXXI organization and its effects on all battlefield operating systems. We then conducted collective training prior to deployment for rotations at CMTC and live fire exercises at Grafenwehr.

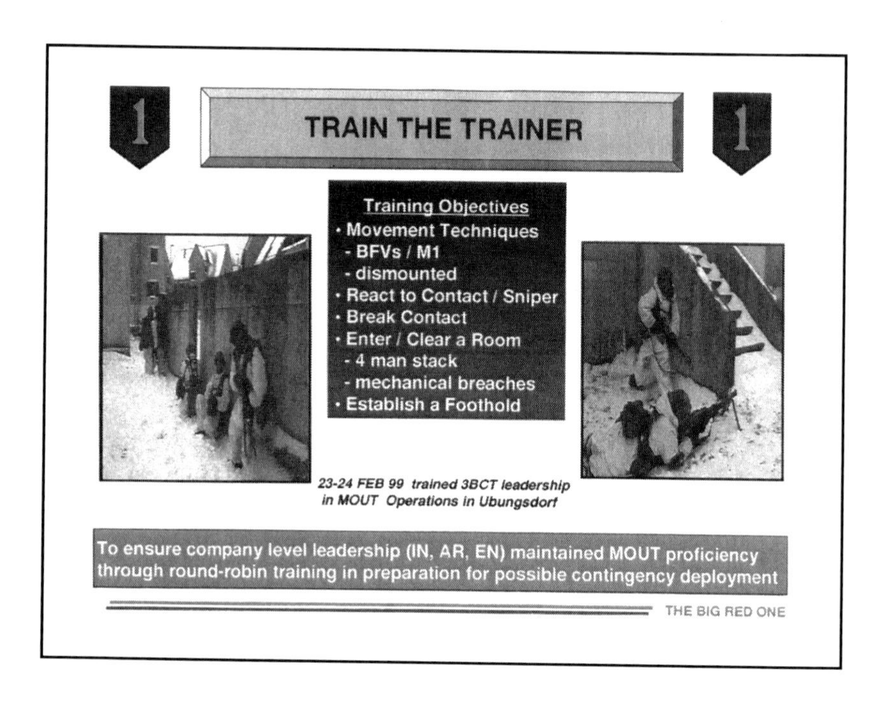

After the leaders were trained, they trained their soldiers. The division ensured sufficient resources were allocated for training at all levels. We had to increase the number of miles we put on our vehicles at the platoon level because previously allocated levels of use were not enough when the demands of MOUT and other combat training were addressed.

M1 UNIQUE CAPABILITIES

- Tanks are the most effective weapon for heavy fire against any structure
- 120mm HE round can penetrate multiple floors and walls, it can open entry ways for the infantry
- 7.62mm COAX and .50 Cal provide additional suppressive fires, though the crew may be exposed to fire
- Tanks are highly effective in a support by fire mode with adequate standoff capability
- Thermal optics can detect heat signatures through walls

THE BIG RED ONE

We analyzed our vehicles' capabilities and vulnerabilities. It was tough to teach tankers how to operate their tanks in and around built-up areas, but it had to be done.

M1 LIMITATIONS

- The M1's max elevation is -10 and +20 degrees. This leaves the tank unable to fire into basements and high level floors:
  - 90 meters away from a building, 9th floor coverage
  - 30 meters away, from a building, 2nd floor coverage
  - 10 meters away, from a building, 1st floor coverage
- When buttoned, tank crews have only limited observation capabilities to the front and sides and no visual capability from top of the tank
- 35ft. dead space around the tank
- Blind spot is created over the back deck due to 0 degree depression and tank must pivot to convert the rear target to a flank target

THE BIG RED ONE

M2 UNIQUE CAPABILITIES

- 25mm and M240 Coax provide sustained, accurate, suppressive fires and counter-fires
- APDS-T can penetrate up to 16 inches of concrete and open entrance routes into buildings for dismounted infantry
- HE is highly effective against earthen / sandbagged structures, penetrating up to 36 inches easily
- TOW missiles can be used against hardened structures and bunkers, penetration is up to 48 inches of reinforced concrete
- BFV armor provides protection for dismounted infantry soldiers to small arms fire

THE BIG RED ONE

M2 LIMITATIONS

- Buttoned, BFV crews have only limited observation capabilities to the front and sides and no visual capability to the top of the vehicle
- Susceptible to AT and RPG fire; must be protected
- In MOUT there are various obstacles to ATGM fires, these include:
  - brush
  - rubble
  - high power lines
  - walls and fences
  - vehicles
- Minimum distances for ATGM (65 meters)

THE BIG RED ONE

M2 LIMITATIONS

- ATGM fire is the least preferred method, due to high cost and small basic load
  - 7 TOWs max per vehicle (2 ready, 5 storage)
  - 3 Dragons max per vehicle (allows only 2 TOWs in ready rack)
- When firing APDS-T there is a 35 degree 100 meter danger zone in front of the BFV due to the discarding SABOT fins
- Sights are susceptible to being destroyed by sniper fire

THE BIG RED ONE

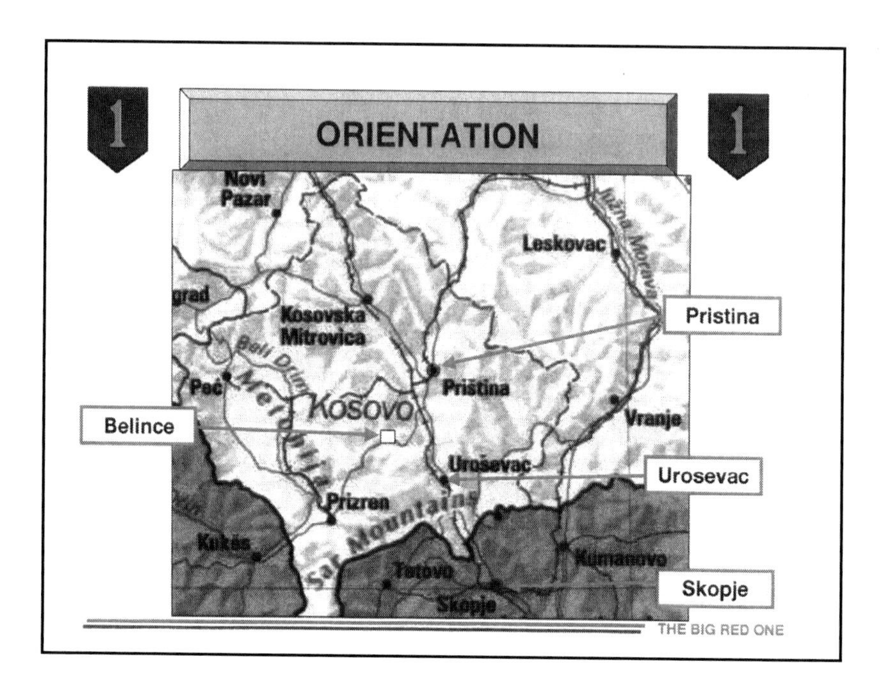

We trained using scenarios involving the terrain in which we would conduct operations if deployed to Kosovo. Training included peace support and combat operations.

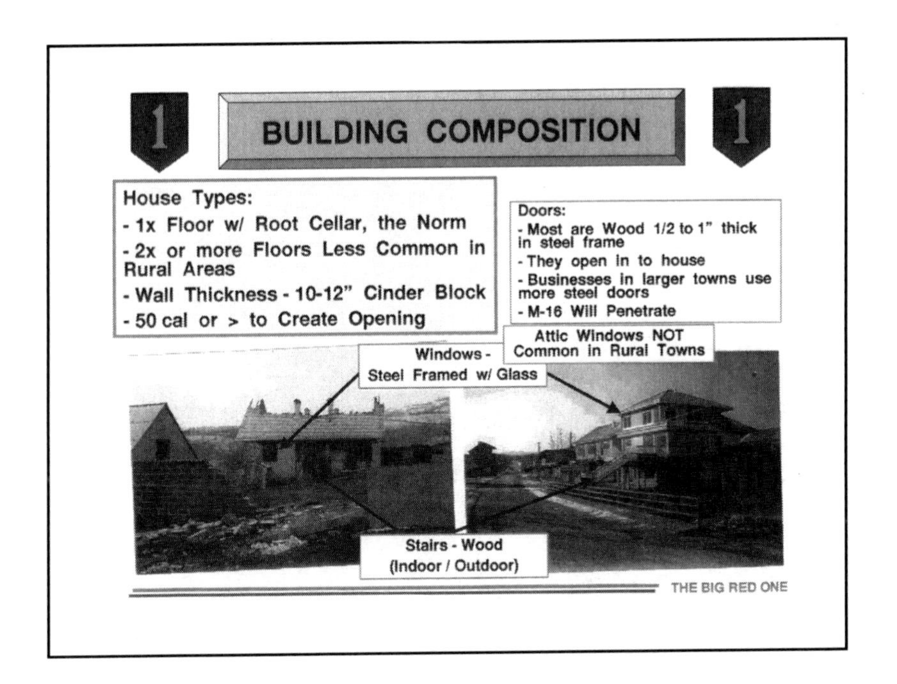

We studied the construction, materials, outside characteristics, and interior designs of Kosovar structures. We analyzed the distances between locations near a village that would offer cover and concealment and determined how we would move or maneuver into the built-up area.

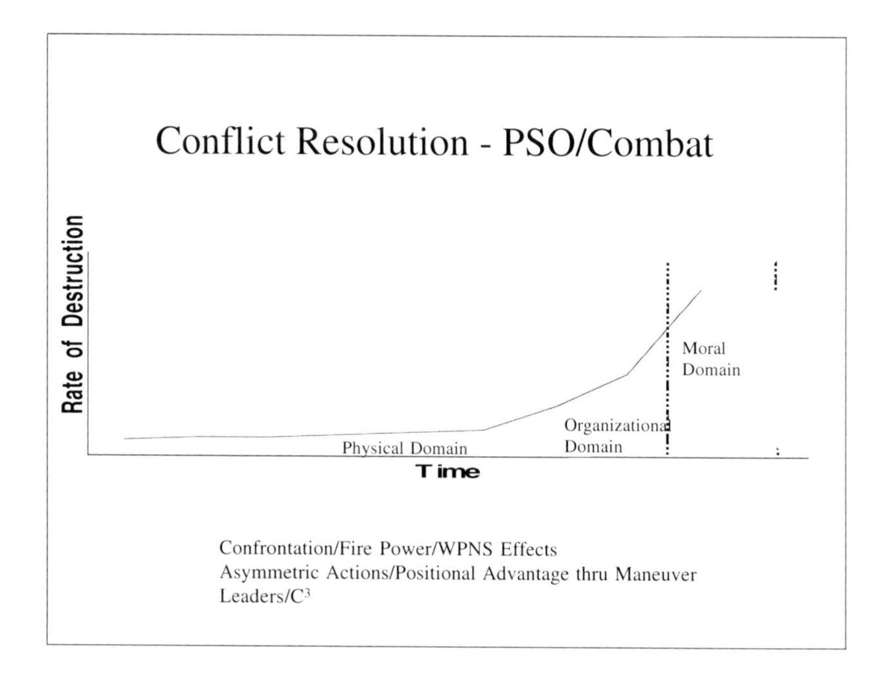

I developed this conflict resolution model using positional advantage and maneuver theory ideas from Dr. Jim Schneider, Liddel Hart, J.F.C. Fuller, and Sun Tzu. If you look at the chart, the vertical line is the rate of destruction of a unit. It applies to both friendly and enemy organizations. The horizontal line is the time it takes to complete the destruction. The model is applicable to urban fighting both in combat and other scenarios. It also applies to stability or support missions, when a unit must deal with a crowd. In the physical domain a force seeks to destroy the enemy and its equipment. The situation is similar when dealing with a crowd. The same principles apply. If you can accelerate the adversary's loss of ability to fight, or conduct a riot, by simultaneously attacking him in the physical, organizational, and moral domains, you have a better chance of winning and keeping your casualties down. This is less costly and faster than taking them on sequentially during which you are likely to lose the battle of time. Based on what we learned from peace support operations in Brcko and during other missions, we need to focus our actions on the right side of this chart.

To do this, you must have good situational awareness, awareness gained through robust reconnaissance using a variety of information systems. The focus in urban environments is on HUMINT. You must understand the terrain and other aspects of the urban environment, such as the population's culture, and those characteristics that might provide opportunities for exploitation. You must see yourself from the adversary's perspective and from that of other relevant groups. And you must see the enemy and noncombatants for what they are, not what you want them to be. You can't play American football on a European soccer field. You must create and shape conditions for operating with a faster decision cycle. You must make the first move rather than react. Your force must seek opportunities to gain positional advantage. You've got to get your forces in the right place at the right time with the right mix during urban ops. You must not be predictable. Your force must use deception and the indirect approach. You must emphasize simultaneity of actions to diffuse the adversary's ability to react. Take actions to degrade his communications and supply operations. Disrupt his movement. In Bosnia the 1st Infantry Division disrupted the enemy's bus movements. Without such transportation he could not conduct riots as planned. Instead of taking him on directly in any one area, we took him on with delaying operations as he attempted to move through the Republic of Srbska. You can also disrupt an adversary's tempo by confronting him with multiple problems simultaneously. Such problems may not be military in character; they might involve seizing his black market money. The bottom line is to maintain your balance while keeping the adversary off-balance. You want to shatter his morale and organizational cohesiveness to impede his ability to operate.

Urban operations are manpower intensive and inevitable. Machines will not dominate urban combat in the near future; the soldier will remain, as always, the primary weapon. He must be trained and ready.

Appendix K

THE URBAN AREA DURING SUPPORT MISSIONS CASE STUDY: MOGADISHU

The Strategic Level

Ambassador Robert B. Oakley
Former Ambassador to Somalia

Briefing to MOUT 2000 Conference

RAND
Santa Monica, California
March 22–23, 2000

Ambassador Robert B. Oakley

Why Mogadishu? So far as I'm concerned, it's worth taking a hard look at the question because it is the clearest recent example of how the United States can either succeed or fail. We started off doing it well, and we ended up doing it badly. When we began, three to five hundred thousand people had died in the previous six to eight months. Another three to five hundred thousand were going to die in the next six months unless something was done. UN operations were clearly not getting on top of the problem. So the president of the United States decided to commit forces based on a recommendation that came, surprisingly enough, from the Joint Staff. It didn't come from the State Department, which was absolutely dumbfounded when the military made the offer.

There are several reasons we should look at this example. Some superb officers in the United States Marine Corps in particular were engaged in the Somalia operation and they have put their lessons learned to very, very good use; the result is that they've risen high in the ranks of their service. There are a couple of other observations from Somalia that are worth noting as well. First is the excessive preoccupation with avoiding casualties.

General Ernst was the first to suffer this "no casualty" edict from the White House, an edict that I think has become very dangerous for U.S. military morale, discipline, and operational effectiveness. It initially came as a result of Somalia. The second consequence of operations in Somalia was the tacit agreement by the administration and the Congress to blame all problems on the United Nations, thereby eroding the potential effectiveness of the UN in places like Bosnia and Kosovo. Republicans, knowing better because they heard testimony from General Garrison and General Montgomery (the U.S. troop commanders in Somalia), understood that U.S. troops were never under UN command. Nonetheless, they found it a convenient weapon with which to beat President Clinton about the head and shoulders. The administration decided not to contest this, so both sides were saying "U.S. troops will never again be under UN command" when, in fact, they never had been during operations in Somalia. This has hurt the United Nations very badly; the United States does not pay its dues and often opposes proposed operations. And it means that the UN can't operate as effectively now as they have in the past. We brought that on ourselves.

As I consider some of the other presentations given here, it seems that the Somalis are somewhat like the Chechens. For centuries they've been waging their own form of guerrilla warfare, mainly cattle raids, and fighting with each other. Nevertheless, out in the scrub brush of Africa, just like in the mountains of the Caucasus, they've learned how to fight and they're willing to do so. And they're not as disorganized as you might think, particularly when in their own environment. You had some pretty capable people out there. You think about Chechens who held senior positions in the Russian Army; in the case of the Somalis, you had Aideed, who had been a very senior general in the Somali army and was Russian trained. He understood a lot and had an effective, disciplined organization. We made the same mistakes the Russians made in underestimating the enemy.

There's a saying about Somalia that goes back centuries: Three things are most important to a Somali: his weapon, his woman, and his camel—and priorities change. Those are the things that are important. If you think the National Rifle Association has a fixation regarding weapons, it's nothing compared to the Somalis. It's part of their manhood. And they learn how to use them. They also learn how to fight. Like the Chechens, if there's nobody else to fight they fight amongst themselves. But if there's a foreigner who comes in, everybody is perfectly happy to fight him and fight even harder because he's from the outside.

But I want to talk about Somalia, particularly how we dealt with the urban environment. We'll look at it from an asymmetrical point of view and talk about how the civilian side can help the military deal with urban problems.

Evolving (Creeping?) Missions

- UNOSOM I: April 24, 1992 (UNSCR 751)
  - Appoints special representative
  - Deploys 50 observers, possibility for force
  - Facilitate cessession of hostilities and promote political settlement
- PROVIDE RELIEF: August 15, 1992 (UNSCR 767)
  - Provide airlift of humanitarian cargo
  - Provide support services
  - Provide internal security for airlift forces
- UNITAF: December 3, 1992 (UNSCR 794)
  - Create secure evironment for delivery of humanitarian assistance in Somalia
  - Use all necessary means to do so
  - Welcome offers of "member states" to establish a force
  - No reference to disarmament, etc.
  - No political objectives

First let's look at the two slides that refer to something you heard Brigadier Bailey address yesterday with respect to Kosovo: evolving or creeping missions. To me, a mission can't creep if it's embodied in a security counsel resolution. What creeps up on you is the gap between what the people at the political level thought was meant when they set the objective and what the people on the ground, who have to carry it out, thought was meant and what they need to do. I would, however, say that missions evolve, and you can see that above. Keep this in mind as we move through the entire presentation.

- UNOSOM II: March 26, 1993 (UNSCR 814)
  - Demand safety of assistance personnel
  - Disarmament (mandatory)
  - Promote political reconciliation
  - Reestablish police and all national and regional institutions
- ADDITIONAL: June 6, 1993 (UNSCR 837)
  - Take all necessary measures vs. those responsible for attack on UN (i.e., Aideed and SNA)
  - Demand disarmament and political reconciliation
- U.S. UNILATERAL DECISION TO WITHDRAW (by March 31, 1994): October 8, 1993
- JTF Somalia: October 13, 1993
  - Provide force protection for U.S. forces
  - Facilitate continued U.S. support of UN ops
  - As required, secure locs to ensure flow of supplies
  - Prepare for withdrawal of U.S. forces

General Ernst can tell you about the last point here, how those four missions were truncated into two. Active missions were cancelled in favor of passive force protection, which translated into meaning that there were to be no casualties until the time we withdrew. So a very effective JTF was sent to Somalia but not really used very much.

Somalia: Interagency Coordination

1 UNITAF MNF:
- interagency task force coordination (State U/S Chair)
- little advanced planning
- experienced civilian and military leader ship (strategi c, operational, and tactical)

2 UNOSOM II:
- No interagency plan or coordination body until Oct 93
- limited U.S. interaction with UN and coalition
- late Oct 93: established ExCom A/Secs reporting to Deputy Cmte. (precur sor to PDD 56 ExCom)

At the same time, we had another problem. We talk a lot in the U.S. government about interagency problems. Well, here you had a wonderful contrast between administrations. An administration came in with no previous international experience at the top. The experienced Bush national security team recognized that the operation in Somalia needed to have a civilian political/humanitarian component as well as a military side. I had worked with General Powell at the NSC, as well as with the State Department; I worked for President Bush when he was an Ambassador to the United Nations and for General Scowcroft, so I was pulled out of retirement to complement General Johnston. Fortunately, both he and I were given superb staffs of people who had experience in Africa, even in the horn of Africa, and in messy situations elsewhere such as Vietnam where I'd served in the embassy, and Lebanon where once again I had served in the embassy. General Tony Zinni and others had significant experience in Vietnam. Bob Johnston had been a battalion commander in Lebanon. We used our knowledge in discussions during the first week to determine the various traps we could expect and how to avoid them. We kept in mind Ambassador Smith-Hempstone's line that "If you liked Beirut, you will love Mogadishu."

There was the same level of expertise at the strategic level. Remember that this was the team that brought you the Weinberger/Powell doctrine, people who had years of experience both separately and as a team. They wanted to make sure that we minimized the chances of hostile action by understanding the situation on the ground while ensuring that we also had public support and plenty of firepower available should it be needed. The assigned mission was a limited one, a clear mission rather than one that was fuzzy and open-ended and that would drag on forever as seems to currently be the case in Bosnia and Kosovo. It was a realistic mission in terms of the situation on the ground.

Force Structure

USMC (I MEF)		16,200
4 infantry battalions	26 heavy lift helicopters	
1 artillery battalion (30x155 mm)	12 medium lift helicopters	
1 tank battalion (–) (31xM1A1)	16 attack helicopters	
1 AMTRACK battalion (–) (68xLVTP-7)	21 utility helicopters	
1 light armor vehicle battalion (28xLAVs)		
Army (10th Mountain Division)		10,200
3 light infantry battalions	30 assault helicopters	
1 artillery battalion (12x105mm)	5 medium lift helicopters	
	8–16 armed light helicopters	
Air Force	tactical airlift squadron	600
Navy	3 amphibious ships	1,550
Special Operations	1 SF battalion	350
TOTAL PERSONNEL		28,900

Plus 10,000 coalition forces from: Italy, France, Belgium, Canada, Australia, Nigeria, Botswana, Morocco, Zimbabwe, Pakistan, Egypt, and other nations.

This is the force that was used to conduct the operation. It was built around a core of two divisions and commanded by General Schwarzkopf's Chief of Staff for Desert Storm, Lieutenant General Bob Johnston. Fortunately several of the marine units had done some training together on the horn of Africa during operations five years before when they had to conduct a search for a missing congressman in Ethiopia, and during an exercise there the previous year. This gave them familiarity with the region and each other even though there was no specific planning done for the kind of operation that was eventually undertaken. CENTCOM was also in charge of the USAF airlift of food from Kenya to Somalia which began in late August 1992 and provided valuable additional experience. We thought this force structure would get the job done and would do it in keeping with the Weinberger/Powell doctrine after that doctrine was adapted so that it applied to peacekeeping or humanitarian operations. You can see that in the concept of operations on the next slide.

UNITAF SOMALIA

Concept of Operations

PHASE

I	ESTABLISH LODGMENT/SECURITY FOR RELIEF OPERATIONS IN MOGADISHU
II	EXPAND SECURITY OPERATIONS TO MAJOR INTERIOR RELIEF CENTERS/LINES OF COMMUNICATION
III	CONTINUE EXPANSION FOR INTERIOR RELIEF CENTERS—"STABILIZATION PHASE"
IV	RELIEF IN PLACE OF U.S. FORCES AND COALITION FORCES BY UN PEACEKEEPING FORCES

The mission was clear and limited. It sounded sort of like the one we heard of yesterday during the discussion of operations in East Timor. There the INTERFET field commanders were given lots of flexibility. They were also chosen for their experience. The orders Bob Johnston and I each received and then gave to our people were the same: get out there and make it work. Form a single State Department-armed forces team without concern for hierarchy. We talked to each other and our staffs talked to each other all the time.

The State Department, for once, responded beautifully. They gave me a deputy and three political officers, all of whom had served in East Africa. Most of them had served in Somalia itself.

We had a seven person team from the Agency for International Development, the Disaster Assistance Response Team (DART). The average age was about 26, but they had two superb older team leaders, one of whom was a Marine Corps Reserve officer. They'd been out there for six months or more working with the various relief organizations and traveling all over the country. They knew it like the back of their hands.

They also knew all of the relief organizations that were out there, and that too helped us a great deal. I received two information officers from the U.S. Information Agency, one of whom had been through the PSYOP program at Ft. Bragg. That was our team.

We were able to pull together and work with the UNITAF Restore Hope headquarters team very, very well. Intelligence is an example. You saw from John Allison's briefing that the military had a considerable number of intelligence units. We also had our own sources. We had Somalis who were former embassy employees. We had our own foreign service officers who spoke the language and were familiar with the country. We put this all together and came up with what the marines call "cultural intelligence." That is, UNITAF developed an understanding of the situation on the ground so when an intel report came in we understood the context in which it fit and how to deal with it. We also worked on PSYOP a lot, both my team and the guys from Ft. Bragg. They worked with the intelligence people to get the substance of our effort right. It was not just projecting a message, but projecting the right message. This helped us a great deal. PSYOP are extremely important, as we shall see later.

UNITAF SOMALIA
HRS's Secured

HRS MOGADISHU—	9 DECEMBER 1992 (MARFOR)
HRS BALEDOGLE—	13 DECEMBER 1992 (MARFOR, CANADA)
HRS BAIDOA—	16 DECEMBER 1992 (MARFOR, FRANCE)
HRS KISIMAYU—	19 DECEMBER 1992 (MARFOR, BELGIUM)
HRS BARDERA—	24 DECEMBER 1992 (MARFOR)
HRS ODDUR—	25 DECEMBER 1992 (FRANCE)
HRS GIALALASSI—	27 DECEMBER 1992 (ITALY)
HRS BELETEUN—	28 DECEMBER 1992 (ARFOR, CANADA)
HRS MARKA—	30 DECEMBER 1992 (ARFOR)

Regarding execution of the operation, we decided to break up the country into humanitarian relief sectors. These are the ones we laid out. I got to Mogadishu ahead of the marines by accident. I went out to Addis Ababu for a UN conference on Somalia. As the marines came out, I realized that if I came back to the United States as planned everybody would be crossing paths, so I went down to see what I could do to prepare the way. The above slide is how the forces were distributed.

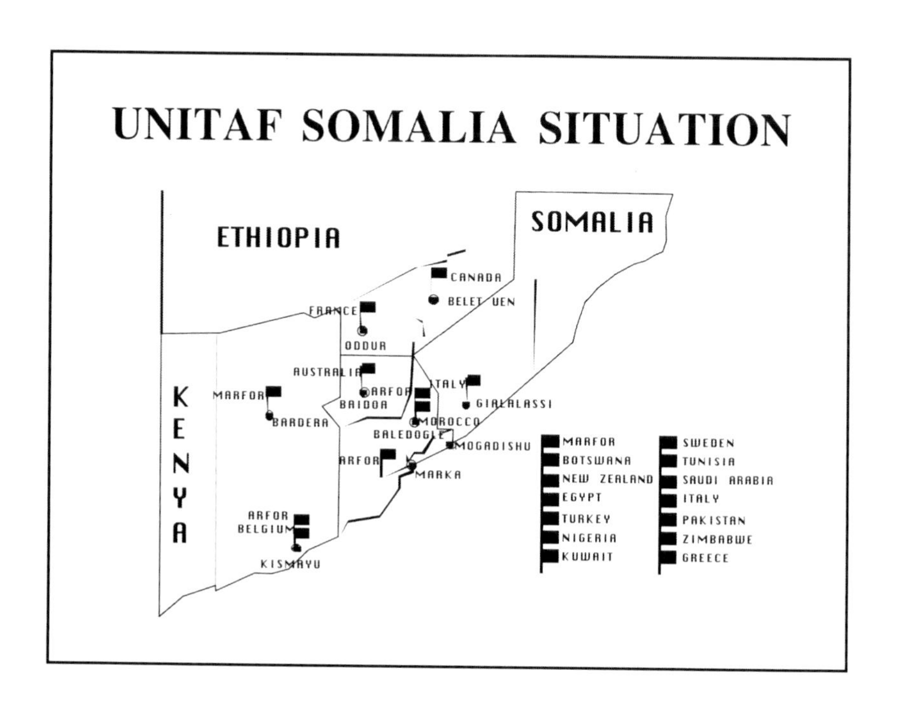

The Bush administration, an experienced team just coming off of Desert Storm, did a good job in establishing cohesion at the strategic level. The concept of operations shown previously was fully understood and supported by the national governments that contributed troops. So we didn't have any problem in that regard. The approach was also understood. Not only was initial contact made with coalition members at the strategic level, but it was maintained throughout the duration of the Bush administration's involvement. This is very, very important to any coalition operation. We heard from Brigadier Bailey yesterday that this was done in Kosovo. It was done in Bosnia; it should be done whether it's a UN operation, a U.S.-led operation, or an Australian-led operation in East Timor, whether it's under NATO, the United Nations, or it's an ad hoc coalition. You have to have coalition cohesion. It starts at the top and you have to keep working it throughout the duration of the operation. You have to watch the operation. It may change on you, in which case you have to go to the top again and make sure everybody's on board. You don't want to have any misunderstandings at the top that are not recognized because, as you'll see, they come right down to bite you at the bottom—at the tactical and operational levels. You also must have a clear command and control system, which we did. We provided each of the other national units with communications and liaison officers to ensure understanding. The following chart shows the C2 structure.

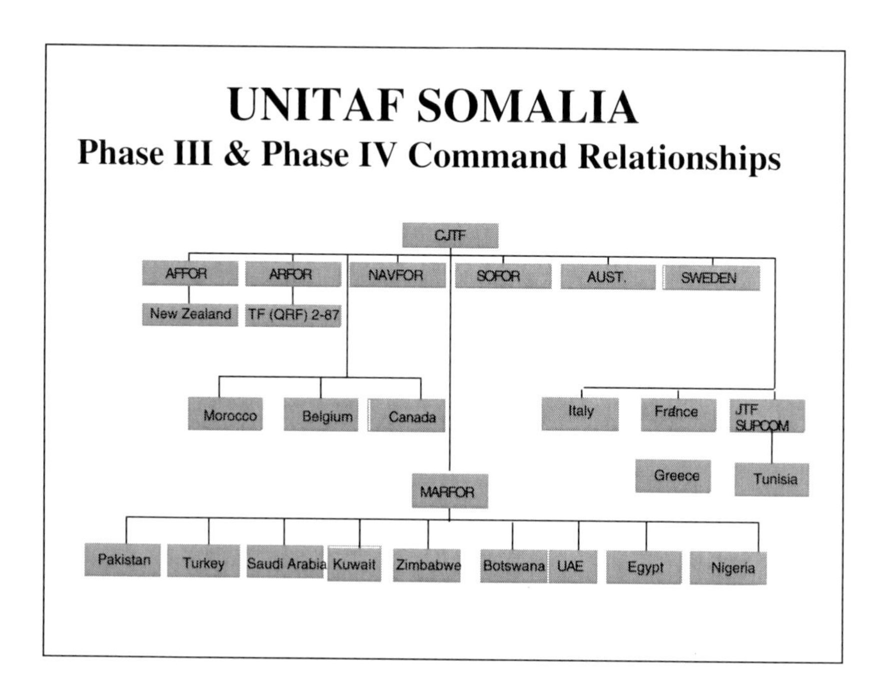

You've seen the dates that the humanitarian relief sectors were occupied. Once we got on the ground, once we got firmly established, we were able to move about twice as fast as had been predicted so that all humanitarian relief sectors had been occupied by the first of January rather than late January. The military took medicine and food each time it moved into the interior. The point that Brigadier Bailey highlighted yesterday about making yourself helpful as well as powerful is very important. After the initial actions in Mogadishu, I was drafted to go first, ahead of the military units, to each of the interior humanitarian relief sectors. Again, the objective was to establish a dialogue with the people in control rather than relying only on firepower. This meant with everybody, not just the military commanders. It meant the political commanders, the clan leaders, the leaders of women's groups, and the religious leaders as well.

Restore Hope in Its Execution

- EYES ON TARGET
- AMBASSADOR OAKLEY/COMMANDER UNITAF RELATIONSHIP
  - Combined Security Committee
- PSYCHOLOGICAL OPERATIONS
- CIVIL MILITARY OPERATIONS CENTER (CMOC)
  - Delivery of Relief Supplies

Eyes on target. This refers to intelligence. We've talked about cultural intelligence; it also meant tactical intelligence. It meant aggressive patrolling. It meant the Fort Apache sort of outpost that John Allison talked about. It meant being out there in the street, knowing what's going on and being seen. It meant moving your patrols and checkpoints all the time so that people weren't sure where you were, weren't sure what size force you had. Colonel Buck Bedard, the "mayor of Mogadishu," was in charge of securing the city. He was a Guiliani sort of mayor and he did a great job in gaining the respect of the Somalis and keeping everybody on their toes, including the other members of the coalition. If he found a particular checkpoint unoccupied or not alert at 3:00 in the morning, he'd go pull the commander of that unit out of bed, saying "Hey, there's something missing out there," and take him to the checkpoint. It made no difference whether the commander was a Frenchman, a Pakistani, or an Egyptian. The Somalis understood that we were there. They also understood what we were trying to do.

The day Bob Johnston arrived, two days after the first marines landed, I was able to get Aideed and Ali Mahdi, the two principal commanders in Mogadishu, together for their first meeting since the civil war had started. Each of them brought 10 or 12 of their lieutenants. After about the fourth hour, Bob was complaining that he had more important things to do than to sit while these guys talked. I said "No, this is the most important thing you've got to do because they have to understand each other and they have to understand us. It's going to make it much less dangerous as we move ahead." At the end of it, the two leaders came up with a seven point communiqué regarding a cease fire in Mogadishu. It covered removing roadblocks, getting the technicals and heavy weapons out of the way, not carrying arms on streets, and several other things that we, with a combination of persuasion and pressure, were actually able to get them to do within about 10 days. So by the tenth day the situation in Mogadishu was calm. You didn't have any shootings; you didn't find any arms being carried on the street; barricades were coming down.

Somali leaders also wanted to set up a standing joint committee to continue such discussions. In part I think they liked the good food that we served them because nobody was getting very much to eat. But seriously, they understood the importance of talking. So every day for the rest of the time UNITAF was there, sometimes a dozen, sometimes 20 people, would meet at my compound to talk things over with each other and with us. I was often there; Bob Johnston was there; Tony Zinni was there. Sometimes we did it with somebody from my staff and one of the colonels, like Chip Gregson or Mike Hagey, but we were there to listen to them talk and to explain to them what we were doing so we didn't have any surprises. This didn't mean they always liked what we were doing. It didn't mean they agreed amongst themselves or with us, but there weren't many surprises.

When it came to operations, the ROE were very, very lenient regarding self-defense or force protection. But you had forward leaning ROE exercised by backward leaning troops because they understood, thanks to the work of the officers and the NCOs, that by showing restraint we were going to be less endangered than by shooting every time we had a chance. The ROE said, "If you think you're threatened, you have a right to shoot." I'd say there were 50 times every day when people could've said, "We're threatened" and then shot for each time they actually used their weapons. The Somalis understood that.

They also understood that we'd react in a very, very strong way and remove the threat immediately when necessary. About the third day, two technicals tried to shoot down a helicopter in the middle of the night. The two technicals were instantly destroyed by rocket fire. I got onto the telephone with Ali Mahdi and Aideed and said, "I assume that these were not your people, and I'd appreciate it if you'd go on the air with your radios and tell everybody that this was not done by you, that you're not having a war with us. Otherwise it's going to be very dangerous for you." They did so.

When General Wilhelm, Commander of the First Marine Division, felt it necessary to take out a compound in the middle of Mogadishu because those inside kept firing at us, it came at a very opportune time because Aideed had convinced the Somalis that he was the hero of the United States. I happened to be in Washington at the time and UN Secretary Boutrous Ghali called and complained to the State Department. I said to tell him to be patient. "If Aideed is smart enough to convince everybody [at the UN conference Boutrous Ghali was chairing] that he's our boy, that's his shrewdness. It's not our fault. You'll soon see what's going to happen to Aideed's closeness with the United States." Because this compound insisted on firing on our marines even though they'd been warned not to do so, they were taken out immediately. This changed the mood of the conference and of Boutrous Ghali toward Aideed and the United States.

Every time we had an incident of this kind we'd go around and talk personally to the faction commander who was responsible and say "Okay, let's go over what happened here." They would come back and say "They weren't my men at all." So we avoided building up the most dangerous thing in operations of this kind—an adversarial mentality, the idea that you're enemies. You can have an uneasy relationship, which we sometimes had. At that point the Somalis were throwing rocks and burning tires; we were using tear gas and pepper spray. Later on during UNOSOM II, they began using command-detonated mines, RPGs, and mortars; U.S. forces were using helicopter gunships. That situation was quite different. But we wanted to maintain it at the original low level. Part of it was due to how we handled the civilian diplomatic and political side. We never broke off the dialogue. No matter how outrageous they might seem, we wanted to maintain the dialogue until and unless we got into a state of actual war. It reduced the chances of greater danger later on.

Psychological Operations

- LEAFLET DROPS CONDUCTED PRIOR TO EACH OPERATION TO SECURE CITIES-TOWNS-CONVOY ROUTES (6,720,000)
- 8 LOUDSPEAKER TEAMS WITH SOMALI LINGUISTS AND TAPES ACCOMPANY MARFOR/ARFOR FORCES ON ALL OPERATIONS
- NEWSPAPER "HOPE" PUBLISHED DAILY FROM 20 DECEMBER 1992, CIRCULATION 20,000 DAILY
- RADIO HOPE BROADCAST TWICE DAILY FROM 20 DECEMBER 1992 IN SOMALI ON AM FREQUENCY, SHORT WAVE PLANNED

Newspaper review	Interviews
News coverage	Short stories
Poetry	Verses from Koran

Psychological operations were extremely important in avoiding conflict. We had excellent support from the Ft. Bragg PSYOP team. We had a radio station and a newspaper, both in the Somali language. Every day we started off with a different verse from the Koran because we didn't want to allow the Islamists to get out of control. We were trying to head off possible threats. We knew that at a minimum the Sudanese and the Libyans were inciting people to attack the United States. We didn't want to let them get out of hand. I met every two weeks with the Islamic Higher Council, a moderate Islamic leadership group, some of whom I'd known 10 years earlier, to explain what we were doing and get their views. Sometimes they had criticisms that caused UNITAF to change its practices in order to accommodate Islamic sensitivities. We put the results of the meetings in the newspaper and on the radio. Somalis like poetry; we had poetry contests.

This was all part of the process of communication and dialogue. We used our radios to overwhelm theirs, particularly Aideed's radio which tended to be inflammatory. Therefore we didn't have to take them down. We'd overwhelm them and they'd try to beat up our paper boys. They tried to get our radio off the air but they couldn't do it. We were just too well organized and strong. It was a very important part of the operation. (There were discussions of taking down his radio later on when the UN had no radio and Aideed was very incendiary. This contributed to the attack upon the Pakistani UN peacekeepers on June 5th.)

UNITAF SOMALIA
Joint Information Bureau Highlights

- Operation Restore Hope was the first major operation since DOD passed its nine principles of media coverage
- Hosted 750 news media representatives (over 95 U.S. media organizations and more than 150 foreign media outlets)
- The JIB wrote more than 250 feature stories
- Published "The Somali Sun Times" weekly for all JTF personnel
- Filed more than 50 radio reports on Armed Forces Radio

Since the objective was to support humanitarian operations while simultaneously conducting these stability missions, we understood that psychological operations in the theater had to be complemented by public affairs operations to communicate news of Somalia back home and to the rest of the world. We worked very, very hard on that. UNITAF had perhaps 35 or 40 people working in their public affairs section. We had two information officers working with them to ensure consistency in our public affairs messages. Everybody understands what's going on all over the world these days, so your message back home has to fit with the message you're giving on the ground. Bob Johnston and I spent an awful lot of time on TV shows. Our staff also spent considerable time working with the media. We didn't try to hide anything. We wanted to be as open as we could. We got pretty good press because we worked hard at it. This bolstered public opinion and congressional support.

UNITAF SOMALIA
Civil-Military Operations Center (CMOC)

- A CENTRALIZED CONTROL POINT THAT GUIDED AND COORDINATED THE EFFORTS OF UNITAF AND THE UNITED NATIONS—NONGOVERNMENTAL ORGANIZATIONS (NGOs)
- RESPONSIBLE FOR DELIVERING OVER 200 MILLION MEALS TO NEEDY SOMALIS
- COORDINATED THE EFFORTS OF MORE THAN 60 DIFFERENT ORGANIZATIONS ENGAGED IN WIDELY DIVERSE OPERATIONS
- ORGANIZED OVER 235 SECURITY CONVOYS AND OVER 130 RELIEF CONVOYS

This was a vital part of the operation. We called the CMOC our "humanitarian operations center" because we wanted to move it out from inside the wire, out of the military headquarters, so that it would be accessible. We therefore put it in UN headquarters. The military side was there as managers and they shared it with the people from the DART (who acted as co-managers). This helped establish credibility in the eyes of the various relief organizations such as the World Food Program, UNICEF, and others. As John Allison talked about yesterday, we had a meeting every day to sort out what we wanted to do during that day and the next, which convoys were going where, what equipment was needed where, etc. It was a humanitarian operation center because all the humanitarian organizations offered their information to help make it work. AID provided some computers to process information and make it easily available. It was where you came for your humanitarian operations information, which was a big inducement for them to cooperate with us. This was also where they could get protection, and sometimes they could get assistance—trucks, for example, or engineers to repair the gates on irrigation canals. The irrigation situation was better than it had been for 10 years. There were also support problems with which the military could help. For example, several major gates from the canals to a nearby river had to be repaired. The army engineers took care of it.

All of this helped establish a cooperative relationship that is not always easy to do with the NGOs. At one point I told NGO representatives that they were as difficult to deal with as the Somali factions.

Nevertheless, we established a cooperative environment and most of the time they would go along. We set up an informal steering committee with representatives of the senior relief organizations and most others would cooperate. There were some that just didn't want to go along, so we let them go off and do their own thing. We were extremely successful overall, and don't forget, these organizations have a lot of influence on the formation of public opinion. They're always talking to the media. They have their own access to Congress, so you want them with you in any kind of operation, whether you're in Bosnia, Kosovo, or somewhere else. They also influenced Somali attitudes.

Laying the Groundwork For UNOSOM II

- STANDUP POLICE FORCE
- CEASE FIRE / DISARMAMENT
- REFUGEE RESETTLEMENT
- ENCOURAGING THE POLITICAL PROCESS

We improvised, and one of the chief objectives of our style of operation was maintaining plenty of flexibility and looking forward. We decided to stand up the Somali police force. We had a big fight with Washington since the State Department and Pentagon said "No, this is mission creep. You can not do it." We countered that it wasn't mission creep; it was force protection. We wanted the Somali police, whom we knew had a good reputation with the Somali people, to be out there on static guard duty guarding the gates. They spoke Somali; we didn't. They understood the body language; we didn't. They can deal with crowds in their own way. They don't have to shoot them or hit them with gun butts, thereby provoking a nasty response. We'd rather let them do the job. We had a long, ongoing fight with Washington.

Somali Police

- NO ADVANCE PLANNING FOR CIVPOL OR LOCAL POLICE
- OBSERVING MOGADISHU SITUATION AMB. OAKLEY SUGGESTED RE-ESTABLISHING THE OLD POLICE FORCE:
    - Force protection
    - Assist UNITAF with static prot ection and traffic
    - Assist UNITAF with patrols
    - Enhance UNITAF knowledge of the environment
  - BUILD POPULAR SUPPORT FOR RESTORE HOPE
    - Old police force regarded favorably by population
    - Not clan-divided or dominated; apolitical
    - Buffer for UNITAF from friction of contact with population
  - BEGIN BUILDING INDIGENOUS CAPABILITIES FOR SELF-GOVERNANCE
- GENERAL JOHNSTON AGREED

The third time we were told "no" I sent a message to Brent Scowcroft, the national security advisor. I said "Look, we're going to get people killed patrolling in dark alleys; the Somali police said they will take the point going into the dark alleys. Why don't you let us use them?" Three hours after that cable got to Washington, our first marine was killed in a dark alley. At that point they said go ahead, but with no material support from Washington and the U.S. government. We were obliged to use whatever resources we could find locally, which we did.

UNITAF SOMALIA
Auxiliary Security Force

- 5,000 uniforms from government of Italy
- 24 vehicles from government of Japan
- 5,000 boots, berets, handcuffs, whistles and batons
- Communications equipment (Motorola), inner-city only
- 4.7 million dollars funded by UN
- Food from World Food Program

We hustled and got it. The Italians had uniforms they'd shipped to the Somali police in 1989 before the civil war started. They were in the port in Mombassa, Kenya. We sent someone down there and found them, brought them back, and the Somali police had uniforms. We did a lot of things in that way. We took some of the technical vehicles that we'd seized, repainted them, and gave them to the police—after removing the weapons. It worked. The old police stations were rehabilitated, partly by the Somali police engineers, partly by our engineers. We had a UNITAF military unit stationed adjacent to each police station in Mogadishu. I think there were 16 of them. This provided them with communications; it provided them with moral support; it provided them with fire support in the event of an attack by one of the more heavily armed factions.

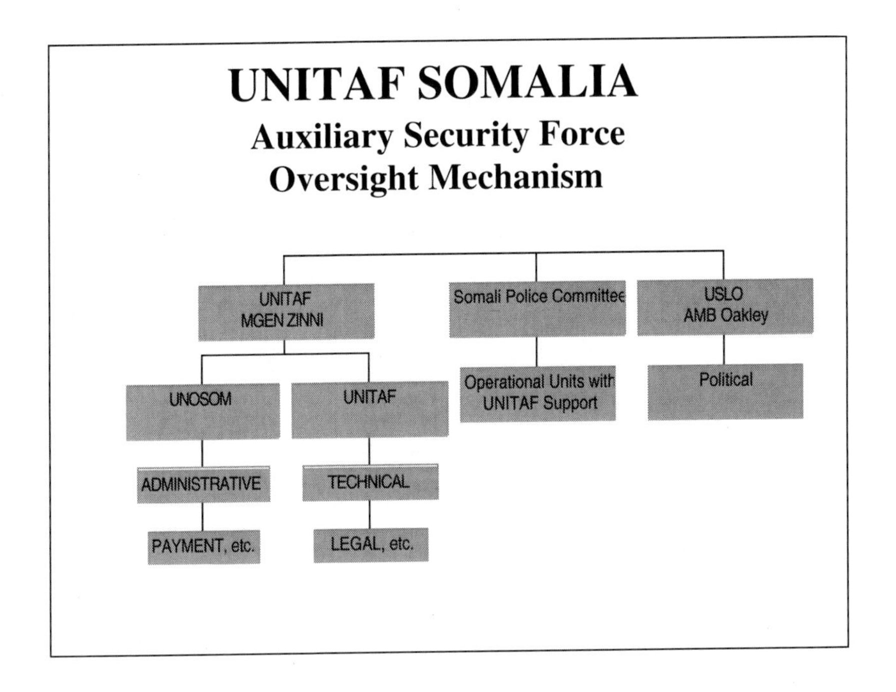

The police didn't want to be under the factions. They wanted to be independent rather than working for a warlord. In the meeting with the senior police, we asked how we were going to run the police force. I told the Somalis that they didn't want to work for UNITAF. This was their country. They were not working for us. So we set up a police committee of 10 former generals and colonels to run the new police force and they reported to the Somali people. Tony Zinni acted as an unofficial advisor of sorts from the military side and I did so from the diplomatic perspective. We had the United Nations provide the administrative support. And this worked—not perfectly, but pretty well.

UNITAF SOMALIA
Auxiliary Police Force

Location	Number	Location	Number
Mogadishu	3,000	Oddur	250
Bardera	60	Baclad	30
Baidoa	75	Bullo Barde	30
Belet Uen	150	Wageed	40
Matabaan	25	Teeglo	30
Gialalassi	50	Afgooye	100
Jowhar	35	Wanla Weyne	50
Marka	65	Kisimayu	200

We had our police in cities throughout the country. I'd say that in five of every six of them, including Mogadishu, the police performed effectively. It was a great help in reducing criminal activity. They were the ones who provided protection to feeding stations. We set up feeding stations in Mogadishu that were providing food to a million people a week, all of them protected by the police working with various UNITAF units; Zimbabwe, Egypt, and Pakistan for example. But these units remained in the background. We never had to use them. The Somali police had little sticks that they used for crowd control. They didn't have to use tear gas, pepper spray, nonlethal weapons, or anything of that kind. They knew how to control crowds and the people obeyed them. If we'd gotten UNITAF into a situation like that, there would've been a huge riot and who knows what would've come of it. A lot of blood would've been shed and a lot of animosity generated. But the Somali police understood. A number of them lost their lives by going after bandits before the United States or others could get there to support them, but they were willing to do that because they felt that they were protecting their country.

Yesterday you heard Brigadier Bailey describe, in a very diplomatic way, some of the problems we're having in Kosovo. You heard about some of the problems that we're having in Bosnia. Many of those problems develop because we are trying to impose our ideas on the people of another nation. We're trying to get them to do things that are unnatural for them. Our approach in Somalia was to allow the people to set up their own organizations and run their own affairs as long as it wasn't done at gunpoint. We had a lot of resistance from Aideed and other clan leaders because they'd been used to doing things at gunpoint.

We had a superb marine colonel who was in charge of the MEU, Greg Newbolt. On D-1 I went into Baidoa, the first big town in the interior, and met with local leaders from the region. There was a Muslim sheik who expressed his concern that the Islamic religion was going to be overwhelmed by foreign presence. I told him he didn't have to worry, that we had a lot of Muslims in the United States and that we understood other religions. The next day, Greg brought an MEU detachment in. The first two people who went up to the sheik were Lebanese Americans. The following day the Catholic Relief Service rebuilt the sheik's mosque. This approach worked well, as did holding town meetings to hear grievances.

At the second town meeting the people complained about a group of bandits that occupied a walled compound on the outskirts of town. The bandits would go out at night raping, pillaging, and robbing even after our arrival; the people asked Greg to take care of them. He looked around and asked whether everybody wanted this to happen. Everybody raised their hands and agreed that was the case. So he sent some of his men over and they pounded on the steel gates and gave the bandits two choices: they could come out peacefully or the marines would come in. If the marines were forced to come in, the bandits were going to come out dead. Using Somali language interpreters whom we recruited from the States, they said "Okay, we'll come out." We found a huge number of weapons: three to four technicals, dozens of RPGs, the whole works. We confiscated it all. The next day Aideed sent one of his lieutenants to see me with a message that asked "could he have his arms in Baidoa back?" I laughed and told him that Aideed's representatives had been in the meeting; they agreed that these guys should be removed. I said "Tell Aideed no way"; the messenger said, "Well, it was worth a try, but we thought you would say no." We developed this type of approach and it helped a great deal, both in this specific instance and in establishing long term credibility. As long as you had the swift and sure use of military force behind you when you needed it, it worked.

We had patrols in the streets at all times. Before the marines arrived, one of my Somali friends said "We won't oppose you, but if we don't like what you're doing, we know how to get you out. We have studied Vietnam and Lebanon." I told him he would never have a chance because they would never, never find us unprepared. And they never did, so after a few initial incidents in Mogadishu, they stopped trying. The U.S. Army and the Belgians had some trouble in a small town called Kismayu, where they were caught by surprise, but there was none in Mogadishu until after the U.S. marines left.

UNITAF SOMALIA

Cease-Fire and Disarmament (Part I)

- 15 January reconciliation meeting established conditions
- UN requested UNITAF help
- Conducted by Somalis, monitored by the UN
- UNITAF/USLO worked with all Somali factions to develop plan

We were even able to get agreement on a voluntary disarmament plan, something the UN Secretary General had wanted from the start. But we did it by negotiation, not trying to impose it by force. It was a voluntary plan to be carried out by the Somalis, not by foreigners taking away Somali guns which would have been an unacceptable humiliation. Since such a huge undertaking could require a year or two and lots of resources to create jobs and to support disarmed men during the transition to another job, UNITAF said they would start the process but that they would be leaving so it must eventually be a UN responsibility. The UN refused, so there was no disarmament.

UNITAF SOMALIA
Cease-Fire and Disarmament (Part II)

- Factions identify assets and personnel
- Cantonment of heavy weapons
- Registration and disarmament of militia members at transmission sites
- Training incentive for militia

[Note: UN did not accept responsibility; UNITAF did not begin implementation.]

General Johnston and his staff did a superb job regarding coalition management. A number of the UN forces, particularly those from Moslem nations, did not want to be seen as fighting other Moslems. They were worried about the reaction back home, so we put them in the airport. We called it UN City. There must have been a dozen national units sitting at the airport that never went out. We tried to put other units where we felt they could be the most helpful and we tried to use them effectively. The Italians were having political problems in Rome because Somalia is a former Italian colony and the United States was getting all the media attention. So we let the Italians run a relief operation. Before their vehicles arrived, we gave them trucks. They used U.S. trucks with Italian flags. They made sure the TV got pictures and it looked good back home.

We had pretty good coalition understanding. As an example, Botswana and Zimbabwe had superb troops. We planned to isolate and search the Bakara market, the place where we later lost the helicopter as described in *Black Hawk Down*. They went through there after a week's training. The coalition took the point with U.S.

Marines in the background. They did an excellent job in confiscating hundreds of weapons without a shot.

Another example: At one point we knew the Nigerians were going to be attacked. We went around the night before and said "You're going to be attacked, so first thing in the morning, bring in everything you've got; we're behind you, but you're the guys responsible for dealing with the attackers." When the attack came, they fired off everything they had. They were firing in a 360-degree arc. They were hitting UNITAF headquarters; they were hitting everything. But the Somalis said, "My God, what the hell did we run into?" and they fell back. Everybody in the coalition felt good about it.

This is the way we worked the coalition on the ground, using field grade liaison officers as well as special forces communicators and communications with each unit to make sure they understood what we were trying to do and to overcome stovepiping between individual humanitarian relief sectors. Every week there would be a meeting of all the national unit commanders at headquarters so as to consistently maintain coalition cohesion.

Stabilization Indicators

INDICATOR	NOVEMBER 92	APRIL 93
BARDERA DEATH RATE PER DAY DUE TO STARVATION	300–325	0–5
GUNSHOT WOUNDS IN MOGADISHU HOSPITALS PER DAY	45–50	0–5
STREET PRICE OF AK-47	$50	$1,000
STREET PRICE OF 50LB SACK OF WHEAT	$100	$7–10
KURTAN POPULATION (NW OF BAIDOA)	10 FAMILIES	800 FAMILIES

Drawdown of U.S. Forces

- 2 HEAVY BRIGADE FORCE: 15,000

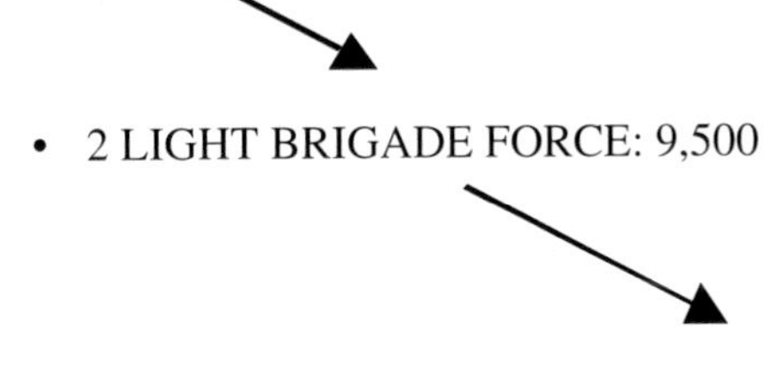

- RESIDUAL FORCE WITH QRF: 4,200

By April of 1993 the situation on the ground looked pretty good. All the HRS were established; food and medical services were being delivered regularly; people were returning to their farms, etc. So the United States was in a position in which it could draw down its forces. On May 4th it finally handed over to UNOSOM II. Let us talk briefly about what happened using slides prepared by the U.S. Force Commander, Major General Tom Montgomery.

SOMALIA AFTER ACTION REVIEW:

Briefing for the Secretary of Defense by U.S. F orces Commander Major General Montgomery
16 June 1994

Purpose

Provide an overview of the Somali After Action Report, primarily oriented on the executive summary and key lessons learned.

Somalia After Action Review (SAAR)

". . . Capture the history of the action, identify appropriate lessons learned, record observations and recommend applicable solutions . . ." —*USFORSOM AFTER ACTION REPORT*

POLICY

LESSON 1— U.S. INTER-AGENCY DECISION PROCESS FOR UNSCR

DISCUSSION:

- Humanitarian focus of mission was *transformed into a broader mission* by SCR 814 with enlarged tasks and responsibilities
- Expanded mission can be traced to a *fundamental disagreement between U.S. and UN on disarmament and geographic coverage*
- The U.S. Interagency level *viewed U.S. role as being limited* and short-term
- While the U.S. fully supported UNOSOM II's mandate, *the objectives in SCR 814 were unrealistic and not capabilities-driven*
- UNSCR objectives were *viewed as UN objectives* for which U.S. would not necessarily be responsible

Somalia After Acti on Review

In the meantime, things had changed within the United States government. The new Clinton administration was operating on two opposing tracks. On the one hand, the U.S. military was sharply reducing its strength. At the same time, at the political level it was demanding execution of expanded missions, missions that were more intrusive and more dangerous. They deliberately introduced what was called "mission creep." The idea was to build a Somali democracy; we were imposing our values on them. It was something they were going to react against. There was also a huge gap in understanding between the United Nations Secretary General and the United States. Boutrous Ghali thought it wise to have a U.S. admiral as his special representative. The Secretary General thought this was the way to get what the United Nations wanted from of the United States. He was assuming that the United States would, in reality, be in charge and responsible even though technically and legally the responsibility was the UN's. The Clinton administration thought this was a good way to support the United Nations and help it succeed while making it a UN rather than a U.S. operation. This way the

United States could get what we wanted out of the United Nations without having to be responsible.

Given the communications gap between our uniformed military and the new Clinton administration, as well as that between the UN and the United States, there were some real problems. The military had not specifically been given the mission of pulling the UN out of serious military difficulties should they occur.

POLICY

LESSON 2—MANNING CRISIS RESPONSE CELLS

DISCUSSION:

- After transition from UNITAF in MAY 1993, *priority of Somalia operations appeared to drop*
- When 5 June attacks occurred, *no high level crisis response cell was available* ; it had been disbanded after transition from UNITAF
- Rush to approve and implement SCR 837 would have *benefited from input/guidance from a crisis response cell* that would have considered the military and political implications of available options

Somalia After Acti on Review

There was the U.S. quick reaction force that was there to work with the UN in responding to particular incidents the UN could not handle. But it was unclear who was supposed to deal with major problems since the UN thought we were in charge and we thought the UN was in charge. Within the U.S. government, the inexperienced national security team hadn't gotten themselves organized; they didn't know how to deal with these things.

The next slide shows the result. This is what happens when you lose coalition cohesion at the strategic level. It immediately impacts operations at the operational and tactical levels because all of a sudden there are no forces. This caused the U.S. quick reaction force to change its mission from support for the UN in case of an emergency to taking the point because there was nobody else around to do so. We did it a little too enthusiastically and a coalition member that was still working with us paid the price.

LESSON 3—PARALLEL LINES OF AUTHORITY & COMMUNICATION

DISCUSSION:

- Coalition cohesion must be gained a nd sustained during Chapter VII operati ons
- Consensus must be sought when there is a major change in mission or operational direction
- Hasty approval of UNSCR 837 did not allow for pri or consultation with contributing nations
- Offensive operations undertaken to meet mandate of UNSCR 837 represented a major shift for UNOSOM II
- Some contingent commanders found it ne cessary to consult with national command authorities before executing orders of the Force Commander
- Unity of effort was adversely affected as coalition partners debated the course of operations both in theater and in New York

Somalia After Action Review

When the Pakistanis were sent to inspect the compound in which Aideed's radio was located, he thought that rumors that his radio station was going to be taken down were correct, so he responded fiercely. At that point the UN declared Aideed to be the enemy and the focus shifted from humanitarian relief to "get Aideed." The coalition suffered badly.

Nor did anyone appreciate the fierceness with which the Somalis would respond to this change in approach, an approach in which the outside world appeared to be imposing decisions on them rather than working with them. At the same time, neither the administration nor the UN understood the dangers of losing coalition cohesion, something that happened pretty quickly after the change in U.S. policy. We were having trouble understanding the change in mission. Not all nations supported the changes. You heard Brigadier Bailey talk about the same thing happening in Kosovo, where a national command authority, whether French, British, American, or otherwise, tells their forces in the theater not to support coalition policy on patrolling or deploying to another sector.

LESSON 4—COMMAND AUTHORITY OF THE FORCE COMMANDER

DISCUSSION:

- Many Force Commander directives providing operational missions to contingent forces were *deferred to national authorities*
- France, Morocco refused further *missions in Mogadishu* following Aideed enclave operation in June
- Italy refused to execute *offensive* missions in July
- Zimbabwe and India refused to *deploy forces into Mogadishu* in September
- Each matter was deferred for resolution *through diplomatic channels* between UNNY and the respective national authority

Somalia After Action Review

We all know what happened. By the time it was over on October 3–4, Aideed proved to be very, very good at urban warfare. He began to generate a lot of support amongst the Somalis—support he didn't have previously. This was the point I made before. They would fight with each other until there was a foreigner to fight. In addition, the UN had no PSYOP, no radio, no newspaper; it stopped meeting the joint security committee; it stopped paying the police and removed their supporting military units. This put UNOSOM in a very weak position and generated hostility amongst the Somalis. Moreover, the Pakistanis, who had taken over in south Mogadishu for the U.S. Marines, stopped patrolling at night, enabling Aideed to regain a lot of control and bring heavy weapons back into town. The UN and the United States were preoccupied with Aideed and scaled back protection of humanitarian operations. The NGOs and the world media turned against the UN and the United States.

On the 12th of July there was a U.S. helicopter gunship attack upon the compound where the Habi Gidr clan, of which Aideed was a member, was having a meeting. They had informed the UN political element as to the purpose of the gathering. The purpose was to see if they could persuade Aideed to leave the country so as to stop the fighting and achieve an accord with the UN and the United States. The military side of the house, namely the U.S. force commander, said "We've got all these clan representatives in one place at one time and we're going to neutralize them for good." This operation was approved by the President of the United States. So five U.S. Army helicopter gunships poured missiles into the compound. By the time this is over, the United States killed 30 to 40 (or more) people; there were no moderates left. Everybody on the Somali side then began to look not to fight the United Nations but rather began asking "How can we kill Americans?" It went downhill from there. The joint chiefs were arguing over the summer for a policy review. There was no policy review forthcoming. Americans were getting picked off in ones and twos. They had to deal with a command-detonated mine here, a mortar attack there. Eventually Task Force Ranger was organized in late August to get Aideed and deal with such problems on a temporary basis until there could be a policy review. Well, while they were thinking about a policy review, lots of things were happening on the ground. We don't need to go into that. The book *Black Hawk Down* describes October 3–4 reasonably well; you have some senior rangers here who know what happened. The result was disastrous. We've seen it again in Grozny. The Somalis knew what they were doing; they said in the beginning "We know what your weakness is." They also understood that our center of gravity in this situation was helicopters. If they could down a helicopter, we'd come to the relief of the crew and they could kill a lot of Americans, getting revenge for the killing of Somalis. They also knew that a volley of RPGs would be as effective against a low flying helicopter as a shotgun against a low flying goose and that the tail rotors were most sensitive. They did not need sophisticated missiles. They were quite successful in what they set out to do.

The result back home was a huge strategic defeat. I see it as analogous to Beirut when the marine barracks was blown up in 1983. You can call it asymmetric warfare or whatever. You can go back to Vietnam and Tet 1968. In all three cases the U.S. body politic, the Congress, and public opinion were not prepared for what happened. Therefore the reaction was "We've lost, let's get out of there. Let's cut our losses and go home." We went home from Beirut. We went home from Somalia. After a while we went home from Vietnam, but Tet was the turning point. The military commanders said in Vietnam "Look how many of the enemy we've killed; this is a victory." I've heard this with respect with what happened in the battle for Mogadishu on the 3rd and 4th of October as well. That may be right; they killed a hell of a lot of Somalis, but on the other hand, you've lost your support at home and therefore you've lost the war. And that's what you have to avoid which is why you have to pay attention to public opinion and to Congress. You have to work with them and keep them informed. Don't get caught by surprise in urban warfare; if you do you may find you have lost the war at the same time that you technically won the battle. Those are the points I wanted to make. During UNOSOM II there was no cultural intelligence, no real understanding of what was going on—of what effect the outsiders were having on the local people. Aideed's PSYOP efforts and public relations campaign directed against the United States was extremely effective. They were reinforced by the reactions of the various relief agencies, the members of which thought that what the United States and UN were doing was insanity. And different countries around the world began saying "We didn't opt in for this, for war with Somalis; we sent forces to help them." Moreover, there had been no meaningful communication with Congress by the administration to tell them how and why the situation and the mission in Somalia had changed so that U.S. forces were in danger. So when the time came there was no more support left in the United States. The whole thing flipped.

After the events of October 3–4 had their decisive impact, President Clinton met for about three hours with Congressional leadership on the 7th of October. He was able to convince them to allow six months for a gradual withdrawal. At the same time he made the decision to send out another joint task force with a lot of firepower. Tony Zinni and I were called back to see what we could do about getting Chief Warrant Officer Durant released by Aideed's forces. We were also told to see if we could arrange a cease fire and try to shift the focus back to a political track. To some degree we were successful at doing so, but we were never able to put the whole thing together. By that time Somalia had come apart again, all the old feuds had been rekindled, and the United States and UN had lost their credibility.

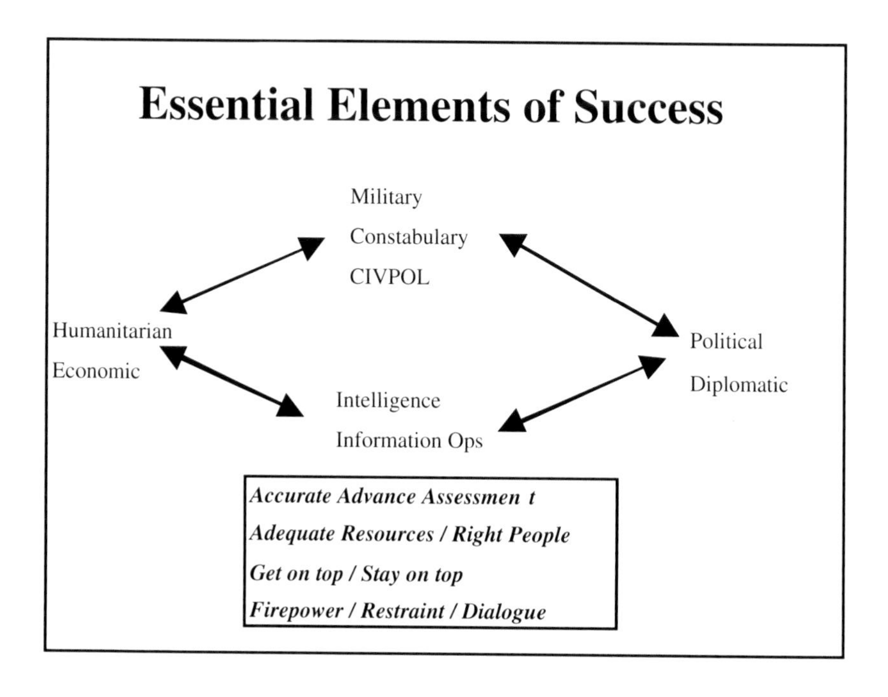

My lessons learned and those of then Major General Zinni end the presentation.

20 Lessons Learned

1. The earlier the involvement, the better the chances of success.
2. Start planning as early as possible and include everyone in the planning process.
3. If possible, make a thorough assessment before deployment.
4. In the planning, determine the center of gravity, the end state, commander's intent, mission analysis, measures of effectiveness, exit strategy, cost capturing procedures, estimated duration, etc.
5. Stay focused on the mission and keep the mission focused. Line up military tasks with political objectives; avoid mission creep; allow for mission shift.
6. Centralize planning and decentralize execution during the operation.
7. Coordinate everything with everybody. Set up the coordination mechanisms.
8. Know the culture and the issues.
9. Start or restart the key institution(s) early.
10. Don't lose the initiative/momentum.
11. Don't make enemies. If you do, don't treat them gently. Avoid mindsets.
12. Seek unity of effort/command. Create the fewest possible seams.
13. Open a dialogue with everyone. Establish a forum for each individual/group involved.
14. Encourage innovation and non-traditional approaches.
15. Personalities often are more important than processes.
16. Be careful who you empower.
17. Decide on the image you want to portray and stay focused on it.
18. Centralize information management.

19. Seek compatibility in coalition operations. Political compatibility, cultural compatibility, and military interoperability are crucial to success.

20. Senior commanders and staffs need the most education and training for non-traditional roles. The troops need awareness and understanding.

Appendix L

THE URBAN AREA DURING SUPPORT MISSIONS CASE STUDY: MOGADISHU

The Operational Level

MG Carl F. Ernst, U.S. Army (ret.)

JOINT TASK FORCE SOMALIA

Carl F. Ernst
Major General, U.S. Army

I'm going to give you a briefing, a sort of after action review, that provides the thoughts of this retired major general. It is not the thoughts of the U.S. Army or of any other agency. I'm going to take you through JTF Somalia—what it did and what it learned. Then I'm going to give you the "so what." The "so what" will help us transition to a discussion of the present.

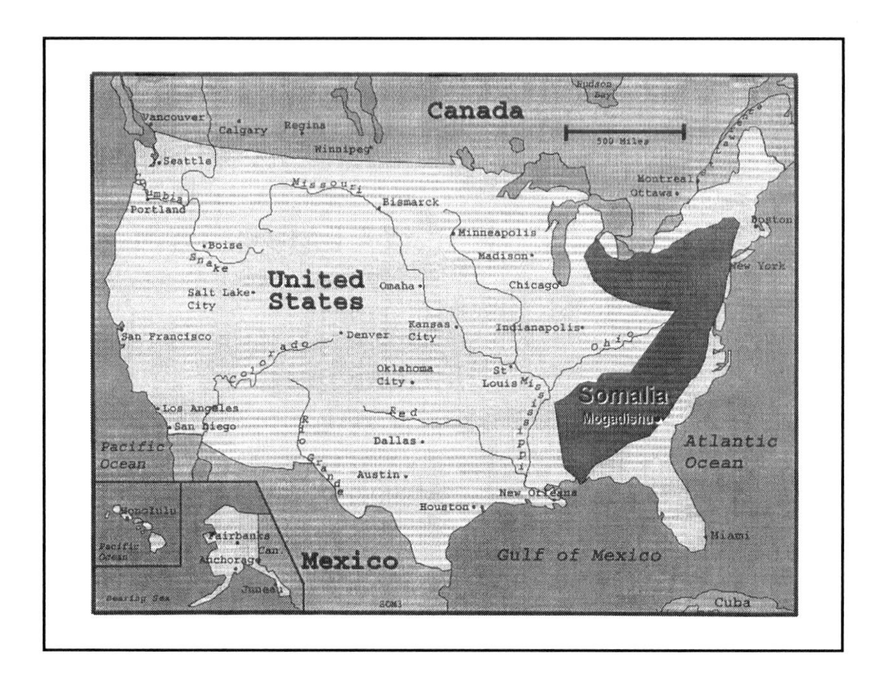

In case you forgot what Somalia looks like, you see it here. There are two things you can take from this slide. Somalia is not exactly a small country. Second, the area of interest in Somalia was such that if you drew a line right across the hook (which was the old colonial dividing line), everything north was fairly stable, and everything south wasn't. The few times that the Somali National Alliance (SNA) tried to go north they basically got their butts kicked and they went back south again. The north was once part of the British empire; the UN area of operation was that south of the line. Most of the focus in Somalia was on Mogadishu, an unfortunate reflection of the lack of both U.S. and UN campaign planning.

JTF Somalia Evolution Chronology

3-4 OCT:	"Ranger Raid"
7 OCT:	Decision to Reinforce/Estab JTF
8 OCT:	CJTF Briefed at CENTCOM
13 OCT:	Initial JTF-HQ Nucleus Arrives in Somalia
15 OCT:	JTF Command Element Arrives in Somalia
17 OCT:	JTF Prepared To Accept Battle Handover
20 OCT:	JTF-Somalia Activated/Accepts Battle Handover, Publishes Basic OPORD
28 OCT:	Executed First Operation
30 OCT:	O/A Draft Campaign Plan Published
8 NOV:	JTF-HQ Staff Fully Manned

Here's a little bit of the chronology involved with the so-called "Ranger Raid" of October 3–4, 1993. Although the rangers' name has become associated with this operation, the more you read about it the more you understand that that wasn't really the case on the ground that day. However, the rangers certainly were a fundamental part of the action and a major part of the joint special forces organization that was there.

The raid began on the afternoon of 3 October and concluded on 4 October; it encompassed nearly 18 hours of close, brutal, and personal urban gun fighting.

On 7 October, three days later, the U.S. national command authority decided to reinforce UN forces in Somalia and establish a JTF. You see the sequence here. My phone rang at 10:00 in the morning on the 8th, a Friday. It was the Army's DCSOPS. Interesting conversation. He says, "Carl, there's going to be an airplane at Langley Air Force Base at 11:00. Be on it." Now, this is 10:00. I said, "Sir, can you tell me where the airplane's going?" He says, "It's going to CENT-

COM and from there you are going to Somalia." It didn't go exactly that way, but you can see that we were moving toward that JTF headquarters by the 8th and 9th. I linked up with my deputy, (now LTG) Pete Pace, and he and I got the CENTCOM brief on Somalia and the CINC's guidance. There was no joint task force staff present because it didn't exist at that point.

We went from there to Ft. Drum, New York on the 10th to meet the nucleus of our staff, the command post (CP) personnel of the 10th Mountain Division. For those of you who are not familiar with light infantry divisions, that's 41 officers and non-commissioned officers. They were the nucleus around which we built the joint staff and it worked out very well indeed.

The USMC's Colonel (now MG) Buck Bedard met us there. The CP, led by the interim chief of staff Colonel (now MG) Evan Gaddis deployed the next day. The command group arrived in Somalia on the 15th. We accepted the battle handover and were ready on the 18th. On October 20, 1993 we published a basic JTF operations plan (OPLAN).

So it all went down pretty quickly, in considerable part due to the planning conducted by the commander, deputy commander, and J3 as they rode on the same airplane for 18 hours with nothing better to do. The unfortunate thing was that we had no communications en route, so we couldn't do the kind of en route planning that we wanted to accomplish.

We executed our first operation on 28 October, a joint amphibious operation. The landing force linked up with an Army light infantry company and special operations task force. It assumed control of both units.

We executed this operation for two reasons: first, to initiate Phase I of our campaign, and second, to test the JTF's ability to plan, execute, and C&C (command and control) a joint operation. We had a pick-up joint task force, a "two-star" level headquarters, but before we tested ourselves we didn't have a clue whether we had our act together or not. We had to determine our immediate readiness as we didn't have a lot of time to get our act together. The SNA was still holding Mr. Durant, our helicopter pilot captured on October 3rd. The special operations task force that had executed the raid on that date was still in Mogadishu. We knew we had to get our act together to take advantage of the forces at hand and address strategic objectives in a timely manner.

On 30 October we published a campaign plan. Up to that point, nobody had a campaign plan for Somalia. Nobody.

The OPLAN had four phases:

- Phase I: Strategic defense, tactical offense—Mogadishu
- Phase II: Tactical defense—Mogadishu, Strategic offense in the rest of Somalia
- Phase III: Strategic and tactical defense
- Phase IV: Withdrawal

JTF — Somalia Mission

Joint Task Force Somalia Provides Force Protection for U.S. Forces in Somalia and Facilitates Continued U.S. Support of UN Operations. As Required, Conducts Operations to Secure Lines of Communications to Ensure the Continued Flow of Supplies. Prepares to Withdraw U.S. Forces.

This was the mission. (By the way, after our initial OPLAN was published we received the JCS plan and the CENTCOM plan.) Everything in the basic OPLAN directly resulted from discussions with the CINC and our getting his guidance and thoughts on how the mission should be approached. We never replaced this original; we instead "fragged" off this plan with the basic plan remaining constant.

There were three specified tasks given to us by the CINC. We assumed a fourth because the President of the United States said that we'd be pulling U.S. forces out on the 31st. We assumed we'd have to do a military withdrawal operation—a major one, an operational level military withdrawal—which we eventually did. The four tasks were therefore:

- Protect U.S. forces
- Support of UN operations
- Secure (read that as "open") lines of communication. (The real problem was that the lines of communication had closed down)

- Withdraw U.S. forces.

The commander's intent was pretty clear, I think. The key to success? Security. We were on the ground and the bad guys had guns; we said, "That's security, not force protection."

The first requirement for any commander is to secure his force because if you don't secure your force you can't accomplish your mission.

That's why we make it a principle of war. On the flip side is the principle of surprise: the reason you secure your force is so that you're not surprised. If we were going to get into a fight, we didn't want it to be rifle against rifle. We made that decision early on. It wasn't going to be M-16 against AK-47. We would use maximum, overwhelming force. We would engage using standoff and precise overwhelming combat power whenever we could. We didn't want to stack our soldiers outside a door; it is best to never have to stack. The best way to clear a room was precision-guided munitions. Now you don't get to use 2000 pound laser-guided bombs a whole lot, but there's a whole menu of things that are available before you have to stack.

Because we were a pick-up team and had units coming from everywhere, we had to learn. We not only had to get the joint staff ready; we had to get the entire joint team ready while we were also conducting active daily operations. There wasn't enough of any one service available to do it otherwise. And there was a lot to be done. We had a good, adequate-sized force and each player was ready, but we were not yet a joint team. We were going to become a team through joint training and rehearsals at all levels, both of which we did from the time we got there until the time we left.

Add this: "Take the moral high ground." There's a Napoleonic maxim that says the moral is to the physical as three is to one. We've known that as a military society for most of the time we've been writing about military societies. But we hadn't figured it out in a place like beautiful downtown Mogadishu. I think—no, I know—our task force was able to do it.

The end state:

- Everybody's secure

- There's freedom of movement
- There is freedom to completely withdraw our force.

Secure also meant that the soldiers felt good about the way we were conducting our missions, that there was no bunker mentality; they weren't hunkered down. If the commander says it's time to move, you move and you're not worried about it. You plan for it, prepare for it, but with a "we're not going to stay in our bases" kind of a mentality. By the way, you may recall that Aideed had made a statement to CNN not long after the 3–4 October events when the U.S. president declared he was sending in a JTF. In a CNN interview, Aideed had stated that U.S. tanks had better stay in their bases and its marines had better stay on their ships. So the gauntlet had been throw down. We decided to accept the challenge.

The Joint Task Force got lucky on this one because we had a great team. Pete Pace; now MG Zaini Smith, U. S. Army; and MG Buck Bedard. Zaini and I had served together in the 82nd Airborne Division. We had also shared another contingency deployment after Hurricane Andrew, so Zaini and I had done this "get there quick" drill before—including a rehearsal for the Haiti operation when I was the assistant division command of the 82nd, so we knew each other. And then, of course, there was MG Buck Bedard, a great marine, so I really had a world class team and some veterans, in the case of Pete and Buck, Somalia veterans as well. Buck and I first worked together in planning for Operation Desert Storm when he was with the I MEF [First Marine Expeditionary Force] and I was with the 3rd Army, so we had some previous experiences on which to build. There's an old adage "I'd rather be lucky than good." In this one we were lucky AND good.

We had a very small staff. The Joint Task Force staff was 146 officers and NCOs. However, through the generosity of the Naval Component Command and Marine Force Component commander, we were beefed up with liaison teams that lived in the joint headquarters. Each team was headed by a colonel or Navy captain who not only augmented our capabilities but also provided the vital immediate access to their headquarters essential when orders had to be transmitted rapidly and clearly and a response of equal quality had to be received in return, a task of no small proportions when dealing with fuzzy command relationships.

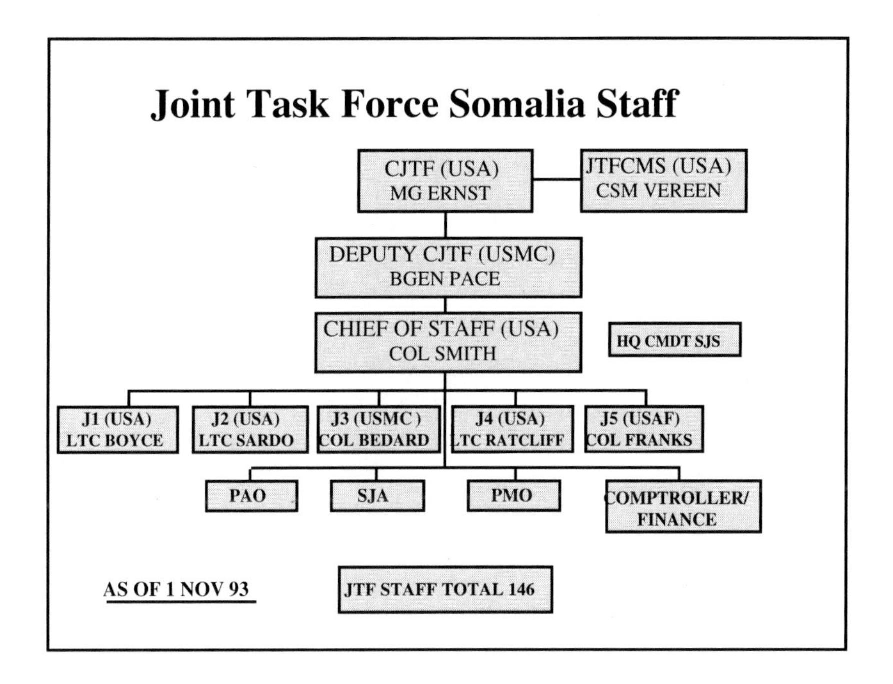

Fuzzy command relationships. This slide shows our command relationships. If Major Ernst gave this solution in any of our staff colleges he would receive an F, but he should get an A+ because it's technically correct. All of the Army force belonged to the JTF with the exception of the FORSCOM logistics element. All the naval forces are TACON when under JTF control. [Remember that TACON meant that it was necessary for the parent headquarters, in this case CENTCOM, to provide the order releasing a force to the JTF. Technically, therefore, we couldn't give planning direction to a force that's to be TACON.] We were once again really lucky. When the Lincoln carrier group arrived, Rear Admiral Jack Dantone was the commander. He is a great naval warrior and naval aviator. He fully understood and agreed to do what we needed on a handshake: "What do you want me to do? We just want to play. We want to scrimmage and we want to play."

Same thing with the Marine component commander. We had two Marine Expeditionary Units (MEUs) that were part of the original task organization. This changed over time, but there was a lot of similarity over time. We had an Army brigade plus two light infantry battalions; the 14th Infantry and 22nd Infantry, 10th Mountain Division; Task Force I-64 Armor (a large mechanized infantry-tank task force); and five other company-sized units, to include a battery of 155mm SP artillery. We had a reinforced corps engineer battalion, aviation battalion, and so forth, so we had a lot of capability. Added to this was a Joint Special Operations Task Force of U.S. Army special forces and USAF special operations aviation. We also added the SEALs to this.

Now look at this. This is U.S. Forces Somalia, which is also the deputy of UNISOM. If you want a Carl Ernst opinion, the things that they talked about eventually leading up to 3–4 October and the idea of U.S. forces under a coalition commander, it's not about the commander. It's about the commander's organization.

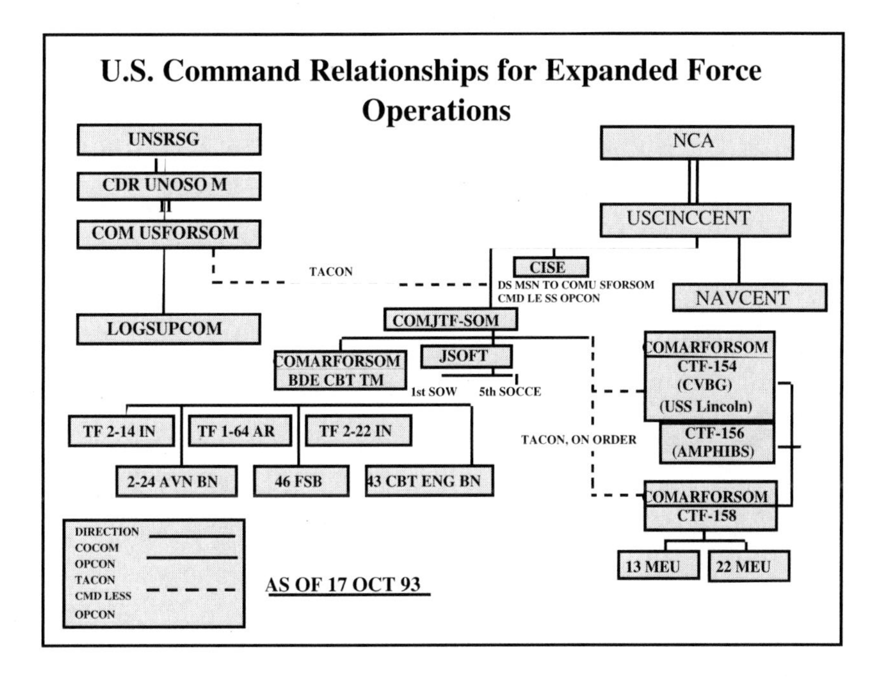

So what makes it a joint headquarters? The ability to do joint planning, preparation, and synchronization during execution. That didn't exist before our arrival. It just didn't exist. We were fortunate that U.S. officers were imbedded in the UN command to assist in these processes. It was in reality a combined staff, not a joint staff, one with very good officers. But in terms of a headquarters capable of planning, preparing, and executing operations, there was no such thing. And that's not good guy/bad guy perspective, it's just that there was nothing there. So when the JTF came, they came with that capability. JTF Somalia worked because everybody came with the attitude that, "We've got a problem. Forget about the command arrangements—let's just figure it out on the ground." Which we did—and it worked very well indeed.

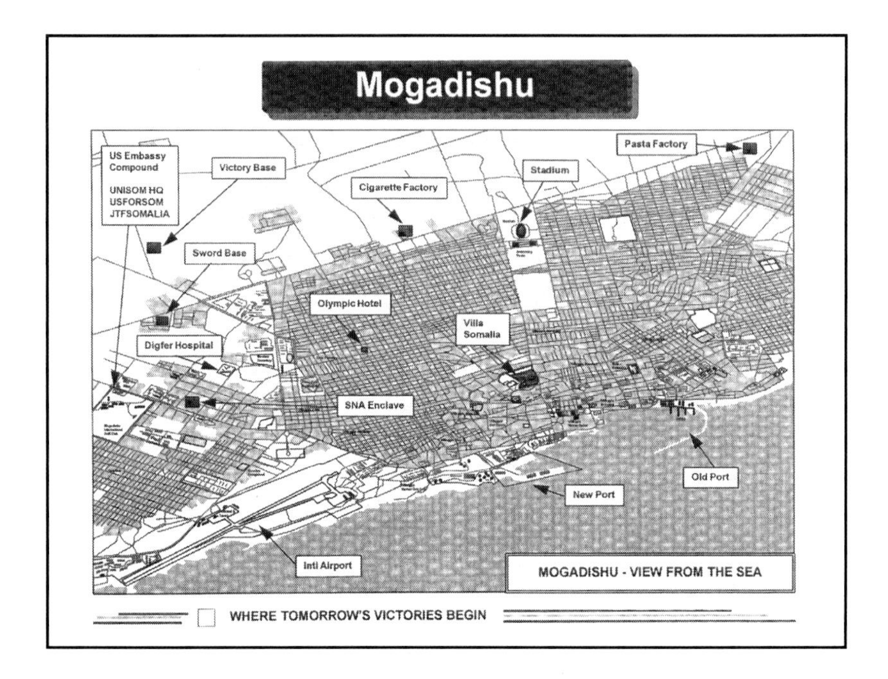

You see beautiful downtown Mogadishu here. This is a map that's very close to what Ambassador Oakley showed during his briefing. Unfortunately, it's the wrong map of Mogadishu—it's the way we had looked at Mogadishu, as if it were viewed from the sea. Now watch this.

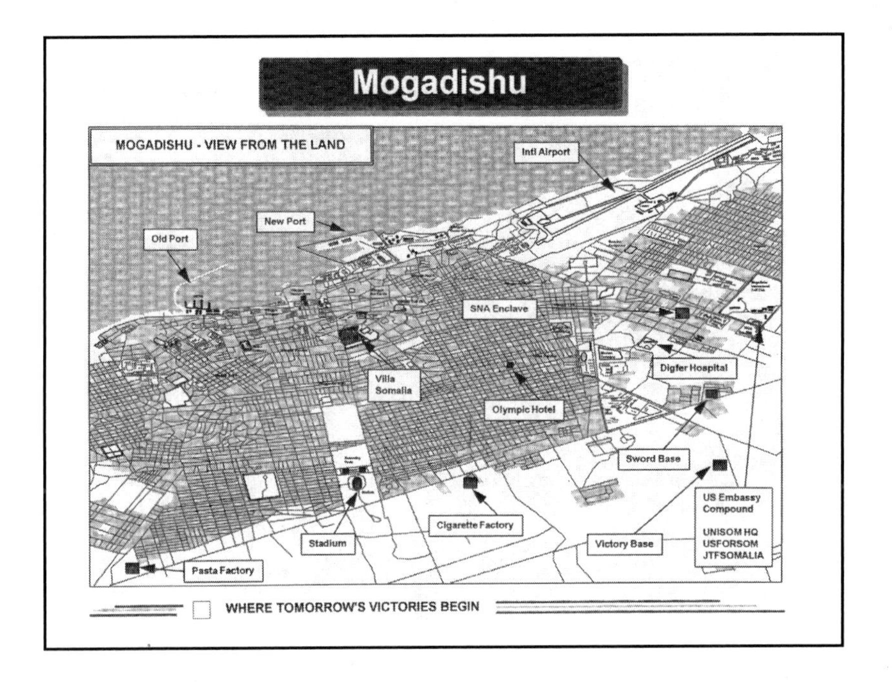

Mogadishu: a view from the land. Now this maneuver wasn't intended to be cute; it was intended to make a point. If you look at the problem from the sea, our positioning within the urban area made sense. But we unnecessarily fixed ourselves by doing so. Our bases were inside the city—right in the middle of the problem. U.S. and UN forces were fixed in a military sense; that is, they lost their freedom of maneuver by their positioning themselves within the city. We figured this out on the airplane on the way over. We decided to change the geometry a bit so as to gain maneuver options and obtain security—all through positioning. We built Victory Base outside Mogadishu for the TF I-64 Armor. I'll come back to that.

Unfortunately, everybody was so wedded to the places they'd been in for almost a year that when the time came to get out of them, we couldn't. The U.S. and UN forces were tied physically and emotionally to these bases. Furthermore, U.S. LOGSUPCOM did not come under the command or OPCON of JTFSOM. By the way, politically we couldn't leave the U.S. Embassy compound because that's where the UN was. I'm not saying we had to get out of all of them, but even our troop and logistics bases were in many cases unwisely positioned. As an example, Sword Base was in the heart of the SNA area, only a couple of blocks away from the Olympic Hotel, the target of the 3–4 October raid. If you walked out the back gate of Sword Base it would probably have taken no more than 10–15 minutes to walk that distance. It's a different view of the world when we talk about how you get freedom of movement and maneuver options in an urban environment. The point here, with this map's view from the land, is that you create installations in the urban area only at locations that you absolutely need—key terrain. The rest go outside unless you intend to clear the city.

JTF Somalia — Campaign Plan

Phase I — Tactical Offense and Operational Defense

Phase II — Tactical Defense and Operational Offense

Phase III — Tactical and Operational Defense

Phase IV — Withdrawal

As I mentioned earlier, we wrote a campaign plan. One of the first guys who received a briefing on the campaign plan was Ambassador Oakley. By the way, if you're going to go on one of these ambiguous operations that has the potential for hostile fire coming your way in a matter of seconds, you hope that the ambassador on the ground, the guy who deals with state and national representatives, to include the executive branch, is a guy like Ambassador Oakley, someone who understands Clausewitzian theory regarding the relationship between diplomatic policy and the military.

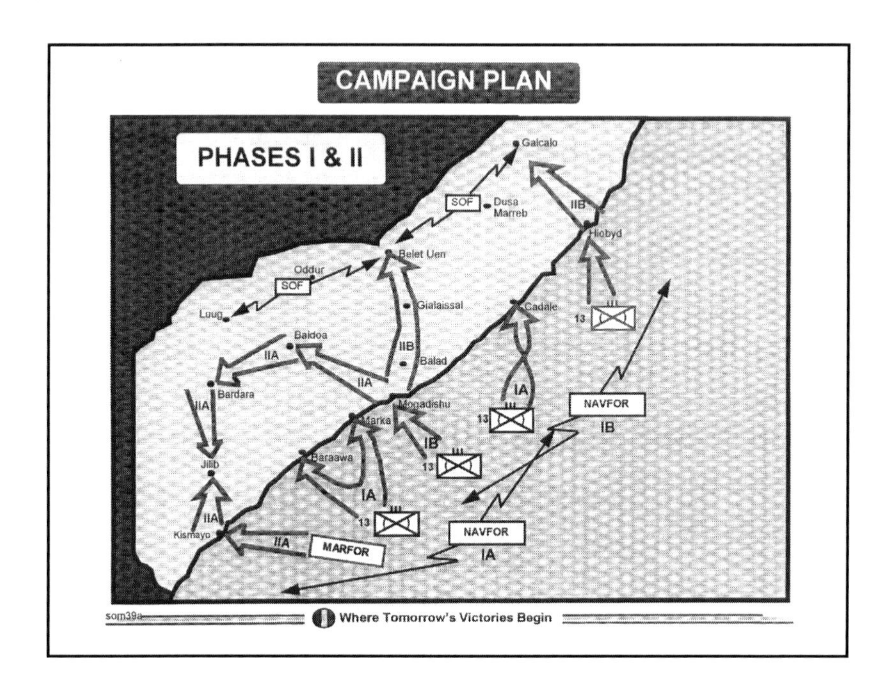

We looked at our operation as one having four phases. The first two were tactical offense and operational defense. Mogadishu was the tactical offense; operational defense was being conducted in the hinterlands of Somalia, which really meant we didn't have enough force to do this all simultaneously, so we had to do this sequentially. Once we fixed the Mogadishu problem we would go on a tactical defense in Mogadishu using UN forces and go on the operational offensive in the hinterlands with JTF forces. This plan presumed a willing UN force. As it turned out, we didn't have that.

The UN was stunned by the events of 3–4 October. Uneven enforcement of the rules of engagement and adherence to UN guidance worsened thereafter because a lot of people were stunned by the brutal fighting that occurred, not just that on October 3–4, 1993, but that which had actually started in June of that year.

The first major event had occurred on the 5th of June. Ambassador Oakley alluded to it during his presentation. The initial JTF had been a U.S.-led coalition commanded by Lieutenant General (LtGen)

Johnston, U.S. Marine Corps. LtGen Johnston's JTF left and turned over operations in Somalia to the UN about 15 May. UNISOM II took over. On 5 June the so-called Pakistani ambush occurred. So called because it was not an ambush. It was the SNA reaction to the UN operation to take out Radio Aideed. The UN decided to go after Aideed. It was similar to the problems we would later confront in Bosnia, Kosovo, and other places. There were no good guys, only bad guys. Some of these groups have been doing this killing for centuries. The Pakistani platoon that went to take out the radio station was slaughtered to a man. Slaughtered. That started it. On the 15th of June the UN conducted the raid on the Habi Gidr leaders' meeting mentioned by Ambassador Oakley—perhaps strategic mistake. It was an armed helicopter attack with consequences we would pay for later.

In Phase 3 (campaign plan) we were to basically go on the defensive throughout Somalia; in Phase 4 we would start the withdrawal. Recall the timeline; on 31 March everybody was to be out of Somalia. We were starting these operations on the first or second day of November, only five months before that date. You have to start the withdrawal process early, we did so in that month of November. The J5 took on the withdrawal plan while the J3 worked on active tactical operations. Our J-5 planners went to 1st Marine Expeditionary Force (I MEF) in December to coordinate and start the planning sequence for the withdrawal because the naval forces, the marines, were absolutely critical to the execution of the withdrawal plan during its last stages. The withdrawal plan itself had four phases and required two MEUs to execute the withdrawal. They had to be task organized for this purpose.

The first operation of Phase 1 (in the campaign, not the withdrawal, plan) was located just south of Mogadishu near Marka. We picked Marka because the SNA had been making movements against the town and because it was a strategic point on the lines of communication to southern Somalia. The SNA needed it in order to move down to Kissmayou. Kissmayou was important to the SNA for economic and military reasons. Aideed needed to defeat a guy named Jess who was as bad as Aideed. The two men hated each other. We went to Marka to do two things: test ourselves and send a message. It was actually a demonstration within a joint amphibious operation.

After that initial action, we started concentrating on Mogadishu. And that's all I'm going to say about the campaign plan.

Capabilities

ARMY OPERATIONS

- Deliberate Attack
- Military Operations on Urbanized Terrain/Close Terrain
- Hasty Attack
- Infiltration Attack
- Area Defense (Point Defense)
- Ambush
- Relief in Place
- Air Assault
- Raids
- Connectivity with Convoys
- Strong Point
- Mobile Strike Force
- Search and Clear Urban Patrols
- Support By Fire
- DART

MARINE OPERATIONS

- Amphibious Raids
- Limited Objective Attacks
- Show Of Force Operations
- -Security Operations
- Tactical Deception Operations
- Fire Support Control
- Relief in Place
- Tactical Recovery of Aircraft, Personnel, and Equipment
- Military Operations on Urbanized Terrain
- In-Extremis Hostage Rescue
- Sniper Teams

One of the things about working with a joint team is that you've got to figure out what each of the players brings to the table. That, I think, is the hardest thing for a joint commander to determine. His staff is there to help him. The structure shown in this slide may not be 100 percent complete and it's certainly not all-inclusive, but it gives you some idea of what was available from the many participants. Look at an Army brigade that includes both heavy and light elements, and then look at a MEU. (By the way there were four of them that came through, and I don't think any two of them were exactly alike. That's not a hit on the Marine Corps; that's just the way they do business. There's actually a lot of good in that flexibility.) There's a lot of common ground between the Army and Marine Corps, but a lot of differences in capabilities that must be understood.

There's a lot in the way of additional capabilities that comes with a carrier group and the amphibious ready group. Depending on who was where we had a carrier group and one or two amphibious ready groups; both of these types of organizations were not identical with others of the same type. One carrier group had a submarine and the other one didn't. All of this was really fun to operate for a joint task force commander. The JSOTF had about half of a special forces battalion.

There was more work for special forces to do than we had special forces to do it, so we cut a deal. Rear Admiral (RADM) Jack Dantone started it. When he left he was replaced on station with America Group (RADM Art Sobrowski commanded America Group) and the arrangement continued. In this arrangement the SEALs were chopped on a handshake to the JSOTF. The SEALs came ashore and reinforced the JSOTF (with one exception that I'll talk about later). Our special operations units were worth their weight in gold and a great combat multiplier. They gave us special recon, special sniper, coalition support team, and special aviation capabilities otherwise unavailable. There's a lot of discussion about intelligence and the ability to get information in an area where the population is composed of a very closed society. This was the most closed society I think I've seen. I've been to Bosnia and other places since, but this society defines closed. Intelligence collection was complicated because we didn't look like them. I don't mean just we the white soldiers; our African American soldiers also didn't look like them. It's how we appear, the way we walk and talk, and so it's very difficult to break into a society that's clan by culture, and then it's sub-clan and they know everybody. HUMINT, special ops, and special recon were very important to us.

Ground Systems

	SYSTEM	O/H	ARMAMENT
Combat Power	M1A1C	30	120mm Main Gun, .50 Cal MG, 7.62 MG
	M2A2	42	25mm Main Gun, TOW, 7.62 MG
	M3A1	6	25mm Main Gun, TOW, 7.62 MG
	M106 (4.2)	6	107mm Mortar
	M109 SP	8	155mm Howitzer, .50 Cal MG
	M198 T	6	155mm Howitzer
	M981 FISTV	5	
	AAV	14	7.62 MG, MK-19/.50 Cal MG
	LAV	14	9x25mm, 7.62 MG, 2xAT, 2 Log, 1 C2
	FAV	14	7.62 MG, MK-19/.50 Cal MG, TOW

	SYSTEM	O/H	ARMAMENT
Combat Multiplier	M9ACE	10	
	DOZER	8	
	AVLB	2	
	SEE	7	
	TPQ-36	3	

U.S. Army equipment
USMC equipment

I'm going to recap the combat power we had available to the JTF: 30 tanks (M1A1 models), 48 Bradley fighting vehicles, mortars, and, this was a key factor, 155mm self-propelled howitzers. The Marines brought M-198 155-mm howitzers. The reason the howitzers were key is because they have a precision fire capability. Q36 [counterfire] radar was very important. The Q36 was digitally linked to 155-mm howitzer units. That meant that with the detection of an enemy round by the Q36, we could put a round in the air going back before the incoming round hit the ground. The Q36 also detected small arms [e.g., rifle or machine gun] fire, so you had to be real careful before responding with artillery.

Aerial Systems

Combat Power

SYSTEM	O/H	ARMAMENT
AH-1F	12	20mm, 2.75 R ockets, TOW
AH-1W	4	20mm, 2.75 and 5" Rockets, TOW, HELLFIRE
UH-1N	3	7.62 MG/2.75 ROCKETS
USAF (SOF)	4	105mm Howit zer, 40mm Bofors, 20 mm
AV-8B	6	Night Cap able (4), 500/200 Lb. Bombs, 20mm
F-14A	14	500/1000/2000 General Purpose Bombs, 20 mm
FA-18C	22	Night Cap able, 500/1000/2000 Laser Guid ed Bombs, 20mm, L2R/Laser, MAVE
A-6E	14	Night Cap able, 500/1000/2000 Laser Guid ed Bombs, L2R/Laser, MAVERIC K

Combat Multiplier

SYSTEM	O/H	ARMAMENT
OH-58A	6	
OH-58D	8	
UH-60L	18	7.62 MG
UH-1V	8	
CH-46E	12	.50 Cal MG
HH-46D	1	
CH-53E	4	.50 Cal MG
KC-135	2	
FA-6B	4	Electronic Countermeasures/J ammer
S-3B	6	ESM Mission Capab le
F-2C	4	C3 Platform
SH-3H	6	7.62 MG

We had a lot of air capability, both fixed and rotary wing aircraft, to include special operations air frames. The OH-58 Delta [helicopter] was probably the most important aircraft that we had during my time in Somalia. This puppy was built as a fire support platform, not as a shooter. Its computer system can store a lot of targets, and it can store an observation point too. If you want to do precision fire, you can load the computer with the aircraft's location and push a button. The little ball atop the helicopter rotates, picks up its target at a designated grid coordinate, and a laser illuminates the target. When this system communicates with other shooters like Marine Hellfire-equipped Cobras, Copperhead-capable artillery units, A-6s, or FA-18s, then you can put a precision round anywhere you want in the city if you get the geometry figured out. There are some capture angle problems at times, but when we had enough firepower located in or around the city, then we could literally shoot a round into any building in Mogadishu from any direction. There was no building in Mogadishu that we couldn't put a precision round into from one of our firing locations. That capability was critical.

I forgot to mention AC-130s. There were unfortunately no AC-130s over Mogadishu on 3 October; the rangers therefore didn't have that asset working for them. We had four AC-130 aircraft in the theater; they were stationed in Mombassa. We could have kept AC-130s overhead 24 hours a day, 7 days a week if we wanted to. Sometimes we'd just keep them overhead and refuel them with KC-135s. It was a tremendous capability.

Intelligence Assets

HUMINT

- National
  - Joint Support Element (JOSE) (CIA, DS to JTF-S)
  - United States Liaison Office (USLO)
- Theater/Joint Task Force
  - Counterintelligence Teams (Marine, Army)
  - Air Force, Office of Special Investigations Teams (OSI)
  - Special Operations Forces (Army/Marine Snipers, Navy SEALs, Army Coalition Support Teams)
  - Civil Affairs Teams

IMINT

- National
  - Secondary Imagery
  - Overhead Imagery

Intelligence: if the United States of America had an intelligence system, we had access to it. We had just about everything you could think of and some things that I'd never heard of. People would come and offer us stuff. We had the ASAS work stations with us, so we had access to national systems in real time.

Intelligence Assets

IMINT (Cont.)

- Theater/Joint Task Force
  - Tactical Air Reconnaissance Pod System (POD)
  - REEF POINT
  - Schweitzer (CIA, DS to JTF-S) (Downlinked, JTF-S HQ)
  - Night Hawk
  - Handheld Aerial and Ground, Still and Video

SIGINT

- National
- Theater/Joint Task Force
  - VOICE/Joint Task Force
  - ELINT (FAST-I)

We also had USN Reefpoint aircraft, USMC UAVs and a quiet, manned surveillance aircraft. All were down-linked to the JTF headquarters. Unfortunately, however, none of these systems had the capability to look inside buildings.

For the most part, all of these systems were useful in establishing movement patterns. When the patterns change, it triggers you to do something. We had surveillance 24 hours a day over Mogadishu and other parts of Somalia that we were interested in.

Targeting. A lot of discussion about targeting. I'm going to distinguish targeting from fire support because there is a difference, a process difference. Everybody knows how you do this, the designation of high payoff targets and the like, so I'm not going to discuss that aspect. Some of the same things that we thought were high priority targets have been mentioned previously. Ambassador Oakley mentioned Aideed's lieutenants. We wanted to know who his lieutenants were by name, where they lived, and what they controlled. Aideed was organized. He had area commanders and then he had specialists. He had an engineer, an example that I'll address later.

Despite good intelligence, targeting was difficult. If a mortar shot at the airfield in Mogadishu, which they had done before our JTF was organized, we had the capability to shoot back at the mortar. Our radar could find it and we could put a round right back on the mortar. But typically when they fired a mortar they would go to an area belonging to another clan, one that might contain noncombatants. So if you fired back you were attacking those who might be supportive to our cause. In our counterfire plan, we weren't really interested in counter mortar or counter battery; we were interested in going after where they stored their mortars, where the ammunition was, and who it was that gave the order to fire the mortar. And so that was part of the process: attack the C2 of the operating systems.

HUMINT was invaluable to this process, not only in identifying potential targets but in helping us decide how to conduct the precision engagement so as to minimize collateral damage. HUMINT and special recon also kept eyes on high priority targets. Special recon could additionally provide us with additional and backup terminal guidance capabilities.

Precision Fire Support Assets

- **Snipers (M21, M24, M40, SR-25, McMillian 300 Win Magnum, Browning BAR 300 Win Magnum, Barrett 50 Caliber, McMillian 50 Caliber)**
- **AC-130 (Air To Ground 105mm, 40mm, 20mm)**
- **Copperhead (Artillery 155mm Laser Guided)**
- **Hellfire (Air to Ground Laser Guided)(Marine Cobra)**
- **TOW Missile (Air/Ground Wire Guided)**
- **Naval Air A6/FA18 Laser Guided Bombs (500, 1000, 2000 Lbs.)**
- **Marine Air AV-8B - Harrier Laser Guided Bombs (500, 1000 Lbs.)**

Precision fires were the fires of choice: ground, air, direct, and indirect. At the lowest level were the snipers. We had all kinds of sniper weapons, basic M-24 bolt guns up to .50-caliber sniper rifles. Special sniping teams went out in the city. By the way, we were not only reinforced with Navy SEALs but also with Marine snipers who brought .50-caliber sniper rifles with them. The conventional Army infantry snipers were given the mission close in around U.S. bases because they were armed with the M-24. The special sniper teams went out in three-man teams and were literally placed throughout the city. If we were interested in a particular area, we would move HUMINT in first, then special reconnaissance, and then special sniping. Rules of engagement were two-way. We tended to say "Sgt. Ernst, here are the rules of engagement—don't violate them." We also made it evident what constituted a hostile act on the part of the Somalis. Hostile acts could be engaged with deadly fire. Our forces didn't have to be threatened. Operating a technical vehicle was a hostile act. Establishing or manning a road block was a hostile act. Carrying a RPG was a hostile act. Possessing any air defense weapon was a hostile act. When we submitted a change to the ROE, it took us about two weeks to get it approved.

Carrying any optic-equipped weapon was a hostile act because it was a sniper rifle. So our snipers could take out anybody committing a hostile act . . . and did. The snipers started cleaning up around U.S. bases first and then started cleaning up the streets around Mogadishu in general. They were the biggest enforcement mechanism we had because, and Ambassador Oakley alluded to it, despite the fanfare of our JTF getting there, the UN and NMCC's will to execute any large-scale operation began to wane.

[Question from audience: Did you try to eliminate Radio Aideed?]

Not only did I not want to take out Radio Aideed, I wanted to help it stay on the air because if we listened to Radio Aideed we got great insights as to what was going on in Mogadishu. One day Radio Aideed put out a broadcast that said, in effect, "Stay away from U.S. bases—don't take these kinds of weapons anywhere in Somalia because the Joint Task Force snipers will kill you." Wow! Bingo!!! That was a good deal for us. By the way, .50-caliber sniper rifles—what a nice piece of kit to take out a technical vehicle or shoot through walls. I mean, just marvelous, just one of the neatest things I've come across lately. Not only can it shoot far, there wasn't a wall in Somalia that puppy couldn't shoot through. A little war story aside: three guys come down the street armed with RPGs and a couple light machine guns. One of our three-man special forces teams (three-man teams because one man acted as observer for the shooters and made sure the areas behind the targets were clear so that we didn't shoot through a target and cause collateral damage) up the street goes BAM, BAM, BAM. The three Somalis are down; the RPGs are left lying in the street. About ten clansmen get behind a wall with light machine guns; they want these RPGs back. The snipers started at the two ends of the wall with their .50 caliber sniper rifles, worked toward the middle, and got everyone behind the wall.

I mentioned to you that we chopped the SEALs to the Joint Special Operations Task Force and they did a marvelous job. The first day the SEALs went to downtown Mogadishu they got into a firefight. They did what they were supposed to do; the snipers dropped some guys and a SNA swarm started. But we just happened to be on the

USS *New Orleans* in a commander's huddle [the commander and his principal staff meeting to discuss an upcoming operation]. I was asked, "What are we going to do?" "Execute the CONPLAN," I replied, and it happened. The joint operations center (JOC) announced REDCON One, the radios lit up, and everybody responded. The petty officer on the roof of the building with the sniper could trigger employment of the assets dedicated to the CONPLAN. He had two AC-130s, four Hellfire-equipped Cobras, and OH-58 Deltas backing him up. They could've put a Copperhead ring around the building if they'd needed to as well as calling on everything else the JTF owned. At that point the armored task force would have been put on REDCON One, standing by for commitment. Everything was on a 15-minute string. A deck alert pair of aircraft immediately appeared overhead with 500-pound laser guided bombs. We had prepared CONPLANs for events that could happen, those that had happened, and had trained down to the point that execution was like a drill for all of them, including what actions might the JOC officer be able to order without any general involvement. When I first got there the sergeants couldn't even use pepper spray. Damn, I mean, the rules of engagement said a man could point a weapon at me, an RPG, and I could shoot him dead, but I wasn't allowed to make the decision to use pepper spray. So we planned, trained, and decentralized execution.

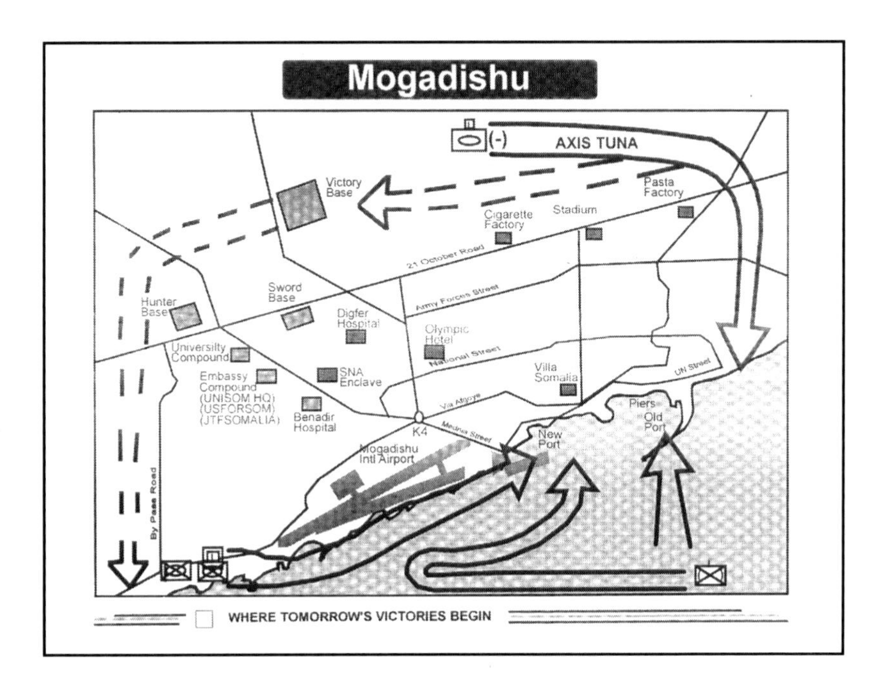

We ran a major operation in Mogadishu about the beginning of November 1993. This operation probably did the most to quiet the neighborhood of any single action we took. It was a joint amphibious operation in which every piece of the Joint Task Force participated. We brought one MEU ashore and then embarked an army mech/tank company team on LCACs that came in on the second wave. A Marine battalion landing team came into the old port area. They established a presence there and operated in the vicinity for three or four days. The Army task force brought in two mech company teams, leaving one at the airport and the other at the new port, then established a reserve force ashore. We positioned artillery throughout the area to provide fire support if needed. That gave us the Copperhead capture angles we needed to shoot into Mogadishu if it became necessary. The artillery fired out to sea for precise registration that night and it was a max sortie day over Mogadishu for carrier and other aircraft, including the AC-130s. We conducted synchronized target engagement employing A-6s, FA-18s, and AC-130s, not over the horizon, but in sight so that everybody could see them. It was just a big firepower demonstration. Once those tanks started operating and the marines came ashore, we picked up the gauntlet. "Okay Mohammed, we got your message, here's ours." That day, Aideed's command net lit up. For the first time we got major intercepts on Aideed's command net. We had put in part of the Marine radio battalion as well, and we had them listening to Aideed's net. So here it is, the first time his radios really lit up. He called his principal lieutenants and asked them what they thought was going on, his effort at a commander's assessment of the situation. There was a UN conference at Addis Ababa scheduled to discuss the Somalia problem. Well, one of his lieutenants says "We think you ought to go to the UN conference" in the clear on the radio. We think we may have influenced that; we don't know for sure, but it was a neat day and it was a lot of fun and it left some JTF pieces in the right places. By the way, I couldn't bring the MEUs ashore unless they were chopped by CENTCOM. But the marines have to bring their gear off the ships periodically to do maintenance and other things. So we put in a maintenance point at the old port. We rotated all the gear there so there was always some maintenance going on... and of course they needed some security so we left a Marine rifle company there as well.

I could also bring one Marine rifle company ashore everyday to train, so we always had at least two Marine rifle companies and an advance battalion command post on shore. We therefore had the basis of a battalion and its fire support shore. (Its artillery was there so I guess that was a violation, but it worked.) It was based on a handshake agreement between warriors who understood the problem and what was needed to solve it.

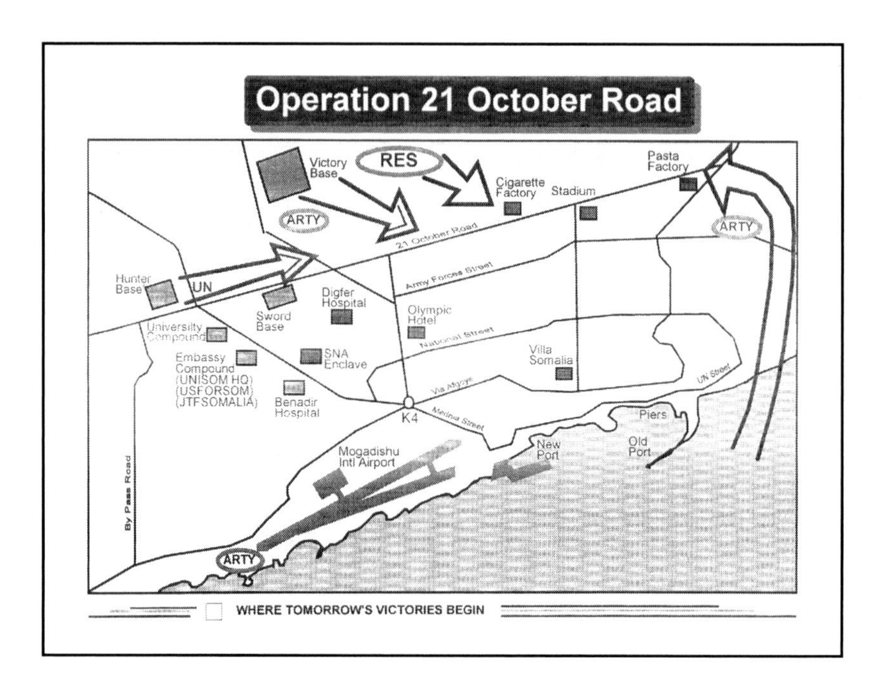

This was the operation that opened the roads. Units were to destroy key buildings if engaged from them. They were to secure key terrain, to include key buildings. These were not random applications of firepower. They were to use precise power. This was Operation 21 October Road.

One of my constraints was that I wasn't supposed to pull point for UN operations. When the UN was supposed an operation, we told them to go up the road and that we would posture ourselves to come in from any other direction if they needed us. If it got hard, they were to just move aside and we'd handle it.

We did not expect a major challenge from Aideed. It was to Aideed's advantage to let us get out of town by the 31st of March. He did not want to challenge us because he knew—Ambassador Oakley told him—that if he challenged the United States his organization would cease to exist as an effective force. Our operations demonstrated that we could deliver on Ambassador Oakley's promise.

Operation Matrix

	D-7/D-6	D-5	D-4	D-3	D-2	D-1
JTF	D-7 - Publish FRAGO (J3 Plans) D-6 - Approved Plan Brief (J3 Plans) COMUSFORSOM - Backbrief security plan (PMO) - Select detainee sites (PMO) - Receive ARFOR/ MARFOR/NAVFOR/ JT SOTF/LSC Backbrief	-Issue NOTAM to airfield (Notice to all Airmen)	- Joint/Combined TEWT (JOC) - COMMEX (JOC/J6) - Airspace Mgmt Mtg (FSE)	- JTF JOC.Mobile TAC conducts rehearsal (JOC) - Air/Fires Rehearsal (Navy/Falcon BDE/FSE/ JSOTF/Marine/LSC/ Medevac	- Joint Illum Rehearsal (Navy/Falcon BDE/ FSE) - Issue notification to close airfield (JOC) - Joint/Combined Rehearsal (JOC) AM dry/PM wet	- B/P second Joint/ Combined Rehearsal (JOC) - Maintenance Standdown (JOC) - Increase base security /no convoys (JOC) - Close airfield to non-tactical fixed wing - Inform Italians of Amphib operations (JOC) - Brief press (PAO)
Sec Force Prot	- Transfer security responsibility from maneuver forces in Sworld & Hunter to LSC (JTF)	- Good MSR lockdown w/ UNOSOM (LSC) - Conduct relief in place of by pass road (LSC)				- Close airfield to non-tactical fixed wing (1800) (JOC) - Close LOCs (1800) (JOC)
Army Forces	D-6 - Back brief COMJTF	- Leader aerial recon - TAC air/OH-58D training - Determine site for PM J/C TEWT - Conduct relief in place of bypass MSR	- Joint/Combined TEWT (AM sandtable/PM walkthru) (JTF JOC) - LNO exchange (Bde) - Unit mission prep - Unit aerial recon - LNO exchange	-ARFOR rehearsal/coord of fires - Unit mission prep - Vehicle TEWT -Air rehearsal (Navy)	US asset/Coalition tark organization TAC air/OH-58D training Obstacle` Unit aerial recon LNO exchange	- Force positioning (JOC)
	(Falcon Bde J3)	(Falcon Bde/Navy)	(Falcon Bde)	(Falcon Bde)	(Falcon Bde)	(Falcon Bde)

Entries also made for MEU ARG, Naval Aviation, SOF, PSYOP, Fire Support, Intel, Eng, CSS, Commo, Cmd, Control, Deception, and CMO

A countdown to D-Day minus 1. These are the events that had to take place during the planning, preparation and rehearsal activities conducted by the joint team. It represents the preparation and rehearsal matrix for one operation.

We told the UN we needed seven days' notice to execute this operation because it would take seven days to get all these rehearsals done. I have talked about the battlefield geometry inherent in properly planning precision fires in an urban area; we actually rehearsed this plan to include that critical aspect. We rehearsed this plan with live fires south of Mogadishu. Every company that was going to participate was part of the live fire exercise. It included precision fires. We fired Copperheads and laser-guided bombs and practiced how we were going to command and control the air space during the operation.

C3 - Command

- **Joint Teambuilding - 2 Parts: The Command & The Joint Staff**
  - **Critical!**
  - **Command Relationships**
  - **Parallel Planning**
  - **Commanders Huddles**
  - **Training Includes Joint**
  - **Drop 2+ Levels – Sanity Check/Feedback**
  - **CSM Part**
- **Overt Jointness (Fairness)**
- **Standards – Discipline (Uniforms, Alcohol, etc.)**
- **Candor**
- **It's All About Trust – Up, Down, All Around**

Command and control: trust was the most important thing in our operations. If we didn't have that, we'd fail. It had to be there at the bottom. The squad leader on the street had to believe in his chain of command. There had to be trust among the coalition forces. One of my jobs was to go out and coordinate with the UN forces on the ground. I went to every UN organization, from Mogadishu to the hinterlands. I didn't just talk to commanders; we went and looked at their positions, which is something the UN didn't do. I saw every position of every coalition force member in Somalia. I wanted the UN to trust us.

C3 - Command (cont.)

- Moral High Ground
  - Us - We're Ready (Heartbeat From a Gunfight)
  - Them - Don't Mess With U.S. Forces
  - Strength - Training OPTEMPO, Standards, High Visibility (Artillery, Air, Ships, Tanks, Victory Base, etc.)
- Strategic To "Tiny Tactical" Daily
- Presence - Troopers

Let's talk about the moral high ground. There are a lot of pieces to this. The troops need to be confident. What we did not want was a bunker mentality. We believed we were a heartbeat away from a gunfight. We wanted the units to believe it and believe they could handle any situation. They could. We weren't going to be in the clans' face, confrontational, but we weren't going to back down. We wanted them to know that if they started something, we were damn sure ready, willing, and able to finish it. Strength, everyday, seven days a week. By the way, the UN conducted operations five days a week. Saturday was a maintenance day and Sunday was a stand down day. Not smart! October 3, 1993 was a Sunday. Leadership and presence? I can't say enough about it. Soldiers need to understand that the chain of command is working for them.

C3 - Control

- Staff – Superb, High Performing
- SAMS Planners – Disciplined Process
  - Includes Joint
  - Includes Commander, DCG, etc.
- USMC Map/Graphic Process
- Navy Air RECCE/Photos
- JOC Drills/Rehearsals Inc With Commands
- Force Tracking – Before, During, After Move
- Force Agility
- CPs
  - Main
  - Mobile
  - Air

The Navy has a photo process that's really great. They can give you pictures and then put graphics on them so that instead of using the maps of Mogadishu we could distribute these photo maps down to the lowest tactical level leaders. Photos of parts of the city with gridlines put on them were invaluable. We could load targets on them, put other graphics on them; we could give them to our pilots. So these photographs were worth their weight in gold and the Navy did this for us whenever we wanted, and it was marvelous. The Marines had a capability to do the same thing on a map or a photo. They could just overprint these things for us, a tremendous capability that we did not have with the Army's forces.

Liaison was important. Special forces coalition teams and fire support teams played a critical role. USAF TACPs withdrew because there was no air war there—big mistake. Marines backfilled with ANGLICO teams attached to army battalions. The Navy sent in a thing called a mobile shore systems terminal, a developmental system that allowed us to access their command and control capability.

C3 - Communications

- Strategic to Tactical
  - Includes MSE/MSRT
  - FM
  - FM/TACSAT Interface
- Tactical Includes Joint - Links to USMC, etc.
- Navy MAST
  - Access
- Commo For Force Tracking
  - "911 Somalia:" FM, TACSAT, "Phone"
- Laser Codes In SOI

Communications: We had just about everything. We brought in vehicle tactical satellite (TACSAT) antennae for convoys. Something that's unique here: we established a 911 Somalia net. Nothing moved unless we could track it for the entire length of its journey. This was determined before movements. Somalia is a big country. There were convoys that had FM radio that would go out of range by the time they cleared the suburbs of Somalia. We changed that. Anywhere in Somalia, if you called and you were in trouble, we could get you. The first response was the deck alert pair of A-6s or FA-18s from the aircraft carrier. The second response was CH-53s carrying Navy SEALs. Next was an air assault by an Army light infantry company with accompanying Cobras. We could get to a convoy almost anywhere. AC-130s would also be a part of that when they were available. It is essential to get laser codes right; we finally did.

CONPLANS

- **Branches Resulted in CONPLANS**
- **CONPLANS Rehearsed Via Continuous Training**
  - **Downed Aircraft**
  - **Convoy Ambush**
  - **Roadblock**
  - **NEO**
  - **VIP Protection**
  - **Sniper Support**
  - **Others**
  - **Response to Coalition *In-extremis***

CONPLANs: I can't overstate the value of CONPLANs. If something had happened before, it could happen again. Things that hadn't happened could. We prepared plans, graphics, overlays, and fire support plans for each one of the possible contingencies. Targets were loaded into the OH-58 Delta computer. I told Ambassador Oakley something for the first time yesterday, something I hadn't told him when he was in Somalia. Aideed would sometimes change meeting locations at the last minute, making it impossible to set up proper security beforehand. Well, nobody moved unless we could follow them, including Ambassador Oakley, though he didn't know it at the time. The most important thing for us was to get to somebody in extremis. One of the questions I asked the CINC: "Okay, boss, you're giving me all this stuff, what's my authority?" He said, "Carl, if there are U.S. forces involved, you can use everything you've got." So we had the full capability of Joint Task Force available to us, but we had to use it judiciously. So we planned, prepared, and executed. We were ready to use precision fires. The objective was to have minimal or no friendly casualties; we also wanted minimal or no collateral damage.

FARs

- 3 Battalion Rule, Then 2 (2 Up/1 Back, 1Up/1 Back) (Depth, Agility)
- Use Full Potential/Combat Power Of Joint Team (Initiative, Agility, Versatility)
- Expect The Worst, Plan For It, Train It
  - "What's The Worst Thing That Can Happen Now?"
- We are a Heartbeat From A Gunfight
- Show Strength - Individual, Collective
- Training, Trust, Teamwork, Troops

FARs: Flat-Assed Rules. For example, the "Three-battalion rule." Always have three battalions: two up and one back; always have a reserve. The reserve will be mobile. That was the tank task force. It would always have at least two tank company teams under its immediate command which meant that we could chop a tank company team to the Marines when they were ashore. I could chop a tank company team to both the Army light battalions and still have a two tank company reserve.

We were a heartbeat away from a gunfight. We lived this.

IEW

- **Strategic To Tactical**
- **IPB Works**
- **Terrain Data vs. Manual**
- **Imagery**
  - **Real Time - Schweitzer, UAV (USMC), Reef Point(USN)**
  - **Photos - TARPS**
- **HUMINT**
  - **CI Teams**
  - **Jose**
  - **Special Recon**
- **Thermal Surveillance Plan (RAS)**
- **AC 130**

IEW: We brought the UCCATS simulation in. It helped us do positioning, plans, and showed us our vulnerabilities. You could do line of sight shots to determine what we could hit from a location . . . or from where the adversary could target us. It helped us position snipers; it helped us determine where were receiving sniper fire from and where our vulnerabilities were.

HUMINT: the most important thing for us was HUMINT. This Joint Operations Support element did that for us—as did special reconnaissance. They told us what was really going on in the city of Mogadishu every day. The Joint Operations Support team worked directly for me and General Pace and General Bedard—that's it. They worked directly for us and we would key the HUMINT effort where it needed to be. The AC-130 was also a great surveillance platform.

Maneuver - Infantry

- Mech Infantry Strength
- Snipers
- "D" Company in Light
- MOUT Training
  - CQC
  - PT
- Train In Vest

Maneuver: Special operations assets were a tremendous combat multiplier. Mechanized infantry and armor has a place in the urban battlefield. It's a big place. We have to have protective vests. We had those in Somalia. The big difference was that we gave them to the riflemen first instead of the VIPs. We had enough to put every infantryman in Somalia in so-called "ranger body armor" before we left.

Maneuver - Armor

- MOUT
  - Commo With Infantry
  - Roadblock Drill
- Convoy Escort
- Position for Mobile Strike Force
- MCOFT With MOUT Include "Dusting Tanks"
- MOUT 120 Round

Armor: Tremendous combat capability. They couldn't build a roadblock in Somalia that this puppy [the M1A1 Abrams tank] couldn't take out. Contrary to what some believe, that tanks wouldn't have made a difference on 3 October, armor would have made a difference. You don't move an armored force in the city unless you prepare a maneuver corridor. That includes a fire support plan. You've got to cover the roofs. We would not put a tank or mechanized vehicle in any street that was not two tanks wide. If it wasn't two tanks wide, we weren't going to get on that street. There are a lot of streets in Somalia that we would have had a hard time getting a HMMWV through. But two tanks wide gave us two tank guns going down the front; if one of them was knocked out, we could still bypass. Tanks led in pairs. Bradleys followed. The reason? The limited elevation for a tank's main gun and coax machine gun. I wasn't worried about that because with the tank-Bradley teams, the tanks were focusing on shooting down the street using crossing fire technique. The Bradleys, with their guns' sixty-degree elevation, could cover the windows and roofs with crossing fire. Rooftops were also covered by aviation that followed behind the armored force. They did not work in front of it and do constant turns around the formation; they remained in echelon behind with fires ready to go down the sides of the formation. By the way, if you were worried about clearing the defile, a Mine Clearing Line Charge (MICLIC) works great. MICLIC was another capability in the formation; it could clear a defile in a nanosecond.

SOF

- **Combat Multiplier**
- **Economy Of Force**
- **Coalition Support Teams**
- **Special Recon & Special Sniper**
- **PSYOPS's and CA**
  - **Prep Of Operational Areas**
  - **Accompany Maneuver**
  - **Accompany Aviation**
- **Responsive, Agile, Low Profile**
- **Joint CSAR**
- **Mix & Match Joint**
  - **Army/Civil Affairs/PSYOPS/Navy Seals/USAF SOA**
- **Learn Capabilities, Give Mission Orders**

SOF: I can't tell you how much it paid off to have these very courageous soldiers, sailors, and marines on the streets in this context. I put Marine three-man teams all over Mogadishu by themselves. They pulled the trigger a lot and provided a great payoff. The tough part was for guys like us to learn their capabilities, give them mission-type orders, and then simply provide support. Regarding snipers: the Marines pride themselves on being snipers and justifiably so. One shot, one kill. Remember the American sniper in *Saving Private Ryan*? We had a kid in Somalia who could have been his twin brother. Little, short kid with a bolt action M-24. This is a new mark for the Marine Corps, it's an Army mark now. Two shots, three kills. Three guys on a technical vehicle, machine guns, traverses toward a U.S. unit. This kid puts out two shots with a bolt-action rifle. Three guys in the back of this technical vehicle are dead. Quantico, you've got a new mark on the wall.

Fire Support

- Aviation and Sniper Direct Fire Support
- Laser Stuff - Geometry
  - Joint Codes
  - Laser Positions (Air and Ground)
  - Position Shooters
- Precision
  - Targeting/Strike
  - Prep
  - Fire Support
  - Counterfire
  - Copperhead
  - OH-58D
- Q-36 Sensor To Shooter

Fire support: I include aviation here. Now my army aviation brethren will say "Aviation is not fire support." It was in Somalia. It's not a hang-up with the Navy and Marine Corps. It's only a hang-up with Army pilots. By the way, most of the time the shooters overwatched the OH-58 Deltas. A Marine Cobra with Hellfire didn't have to be as far forward as the OH-58; he could shoot from way back. So we didn't have to over fly the city to shoot into the city; it was a matter of planning. Further, only the aviation brigade commander could approve transiting the city or any altitude deviations. When planning to employ aviation, it is essential to think through the suppression of enemy air defenses (SEAD), to include small arms and RPGs.

Aviation

- **Support By Fire To Platoon Level**
- **Flight Corridors**
- **Altitude Restrictions**
- **Emergency HLZs**
- **Downed Aircraft CONPLAN**
  - **DART**
  - **CSAR (Joint)**
- **OH-58D!**
- **Airborne C&C**
- **RSTA Includes Joint**
  - **USMC Cobra/Hellfire**
  - **USN A-6/FA-18**
  - **USAF AC-130**
  - **Schweitzer**

Other aviation considerations: With regard to downed aircraft, the first response for any aircraft going down should be to seek a pre-planned emergency landing zone (LZ). These were loaded into our computers. The first response should an aircraft have gone down was with the SEALs unless the airframe happened to be very close to a friendly position. We planned to put a friendly fire envelope around downed aircraft as part of the recovery operation.

A2C2

- Big Challenge in Joint, Precise, MOUT
- Geometry Problem
  - Compressed Battlefield
  - Compressed Airspace

Airspace Command and Control (A2C2): Big problem. This was a real challenge as we frequently wanted to put a lot in the way of air assets into a given airspace. The carrier-based E2C was a tremendous capability in this regard. It was like a baby AWACS [Airborne Warning and Control System]. It became our primary means to command and control the airspace. The backup to the E2C, and everything had to have a backup, was an Aegis cruiser. It couldn't do the job quite as well, but it could perform it if necessary.

Mobility/CounterMobility/Survivability

- Roadblock Drill
- MICLIC
- Survivability
  - "Base" Design
  - Wire Obstacles
  - Bunkers
  - Fields of Fire
  - Search Areas
    = Personnel
    = Vehicles
  - MP Type Searches
    = WANS
    = Dogs
    = Random, Roving

MICLIC: We built an urban area for gunnery and maneuver rehearsals. Tank, mechanized infantry, and USMC LAVs actually practiced gunnery techniques to move through these streets. This was all live fire.

CSS

- No Separate Log Commands
- Friction - Mission vs. Security
- Fighting Skills
  - Leaders
  - Troops
- Force Tracking

CSS: Fighting skills are important to CSS folks because you're not always going have the luxury of grunts [infantrymen] and tanks to guard CSS folks. CSS units must train on combat skills so as to be able to provide their own security. Rear operations doctrine provides the TTP required for CSS units in urban operations.

Force tracking: As I've said, nothing moved unless we could track it and get to it if it needed help. We only moved in convoys, and our convoys had to have C2 and escorting vehicles in the front and rear. Pre-movement training in ambush reaction was also required.

Force Protection

Four Components of Protection (FM 100-5, 1993)

1. **OPSEC And Deception**
2. **Keep Soldiers Healthy; Maintain Their Fighting Morale**
3. **Safety**
4. **Avoid Fratricide**

Force protection: This is an overused, abused term. It means everything from take your malaria pill today to full up gun fighting in Mogadishu. That's a pretty big umbrella to stick over a term. We used it in application to the defense of bases and security, both of which have a robust body of doctrine.

Force Protection Observations

- **"Force Protection" Mission = Defense and Security Operations**
- **Force Tracking Critical**
  - **All Movement**
  - **"911 Somalia"**
- **Secure Everything**
  - **Bases**
  - **MSRs**
  - **Convoys**

We dropped the term "reaction force" and used the term "reserve." The word "react" is one of the things that sends chills up and down my spine because react is the worst form of hasty, and hasty is not the kind of thing that a professional military force ought to be doing. We ought to expunge it from our lexicon. We ought to create reserves, mobile reserves, mobile combined arms reserves. The difference is that a reserve acts, it doesn't react, and it acts based on branches to plans, branches that are rehearsed. That's a tremendous difference, not a subtle difference.

Operations Other Than War

- **Terminology Misleading**
- **OOTW = Gunfighting = Combat**
  - **Mental Preparation for Combat**
  - **Physical Preparation for Combat**

Another troublesome term: MOOTW. The army has since changed the word because we didn't like it. I was one of the guys who went on a crusade to change it. Now we call it "stability operations," but the idea is about the same. Read this as "stability operation"; in Somalia it equaled gun fighting and combat. And I would submit to you that it means the same thing in Kosovo. If you're in a hostile fire zone you've got to be mentally and physically prepared to go from not fighting to fighting and back on a second's notice. It's the three block war in another context. Mental preparation of the force has to lead. Physical preparation follows the mental preparation and the acceptance of the ever-present threat of combat.

Operations Other Than War

Opinion From The Field

"Peace enforcement operations are combat operations normally executed under the constraints imposed by the rules of engagement (ROE)"

— Platoon Sergeant, 2nd Battalion, 14th Infantry Regiment

A sergeant in the 14th Infantry summed it up best: "stability operations" are simply combat operations exercised under the greater constraints imposed by the rules of engagement. If you can't figure it out: If you're drawing hostile fire pay, it's combat.

If you don't believe you're a heartbeat away from gunfire it won't work. Everything's executed as a combat operation. Nothing moves unless it's tracked. As I said, nothing moves unless it's part of a convoy. A convoy consists of military escort and whatever you want to move. It's real simple.

If you have time, you can train for any operation. If you don't have time, train for war and you can adjust for the difference. I really believe that. It's worked for the Marine Corps a long time and it certainly has worked for us recently.

Conclusions: What's the "so what?" This is what I believe after my years as a soldier: We've got to be joint. We have to get our command and control in place. We need to operate as a combined arms and joint team, a team that includes special operations forces under the joint command. We have to conduct mission rehearsals, to include rehearsals using digital capabilities like digital maps. We need to get the colonels and the generals trained properly. It's guys like Carl Ernst that are the target audience.

We must have unity of command. If it's not under the command of the warfighting commander it shouldn't be there in the first place. If it moves into the neighborhood, it ought to be under the command of the guy who's responsible for the ground in the neighborhood.

IPB: This is really important. There are no good TTP for MOUT IPB. You have to interpret and modify the current IPB doctrine. But the same principles apply. Key and decisive terrain: you've got to identify what constitutes the critical terrain, whether it's buildings, areas, facilities, or corridors.

As I mentioned, having digital terrain is an absolute must if you want to do proper commander training; it is also essential for detailed planning.

If an asset doesn't have a real reason to be in the city, try and keep it outside the city. Reconnaissance belongs in there, surveillance belongs in there, special operating forces belong in there. If you think you need to retain key and decisive terrain, retain it. But if you retain key and decisive terrain, you've got to sustain the force that's there.

Multiple corridors: Don't go anywhere if you've only got one way in and out. If you go in one way, you want another force to come in another way, and ideally a third way if the situation goes to hell in a handbasket as it did on 3 October. So nothing goes up a one way street if it would have to make a U-turn should the route be blocked.

A commander has three basic options when it comes to positioning forces for urban operations:

- Outside-In: minimum footprint in;
- Inside-In: maximum inside footprint (for example, Mogadishu pre-October 1993);
- Inside-Out: occupy and clear city, defend it from entry.

Light/Heavy is really important. Set recon surveillance first. Don't move down a corridor if you don't have eyes in the street. I don't want to clear a defile if I don't have to, but I want to know before whether I have to do it or not. If an aircraft goes down in Mogadishu, I don't want to have infantry walk down the street to get to the aircraft. That's what happened on October 3, 1993. I'd like to get an armored force in that corridor and have them move at about 25 miles per hour. And never slow down, even for roadblocks. By the way, an M-1 tank doesn't have to slow down for a roadblock. Unless they put a tank in the roadblock; then they might have to slow a bit.

Targeting is important. Fire support planning is critical, and there is a difference between the two. When I'm talking about corridors, I'm talking about identifying potential targets along the route, some for air, some for ground assets. Otherwise nothing moves. This includes direct and indirect aerial fires. Position snipers before you move as well. You want controlled effects; precision fires give you options for controlled effects.

Mobility—Counter-Mobility—Survivability. You can position bases in a way that increases security. Victory Base had clear fields of fire; you couldn't get within 300 meters of it without being seen. It was outside of the city and outside of the city's mortar range. And we didn't put any lights around Victory Base; we wanted it to be dark because we had the thermal stuff. We had night vision goggles. The first night we were in Victory Base some SNAs tried to play games with us. A company team of Bradleys and tanks flushed them and ran three guys down in the bush. It was a major statement. The SNA never came back again.

Force protection equals security, defense, and the warrior ethos. If you look like a professional force folks are less likely to challenge you. That's taking the moral high ground. One of our agents brought us an SNA survey in December 1993 that compared UN forces and JTF forces. It basically rated the quality of the various forces' force protection, who the SNA thought had their stuff together. We actually had the document. It basically said don't mess with the joint task forces. The UN forces were considered weak. Our troops looked good and they acted like professionals. We had seized the moral high ground.

Counter air defense: AAA and RPGs. If you go to JSEAD doctrine you'd have a hard time handling what I'm talking about. You've got to sort of figure this out on your own. It would be helpful if we figured it out in our doctrine. You do want to use your air and your helos. You can use them in a city, but you've got to figure it out and you've got to get everything working to do that. The idea that we don't use helicopters in the city is wrong. You want to be able to see the airspace. You want to visualize the situation, to include the geometry problem, and put that in your command and control system.

Logistics: This doctrine works real well. We ought to use it.

If you look at the art and science of military operations, urban operations have all the art encompassed by other operations, but in complex terrain I think the science gets more difficult. You've got all those things about leading soldiers, sailors, airmen, and marines that is the art side, but the science gets to be real hard because it's a tremendous geometry problem. It's three dimensional, a horizontal, vertical, and depth problem. You add interior volume and then you've got an even more complicated problem. Momentum and mass are critical. Speed is relative. The moral high ground was one of the key elements in momentum. Once we gained momentum we never stopped until the time withdrawal was completed. Leadership—all leaders have got to be qualified, squad through JTF. Leadership—I can't overestimate its importance.

The urban canyon analogy: The problem with urban "ranges" is that every building is separate, and it's got four "forward" slopes, a top, and it probably has a basement. And it's got interior volume. And by the way, the buildings are not joined together like the terrain that defines real canyons. Each building stands alone. The enemy can use it for covered and concealed movement. It provides an adversary any number of ways to shoot at you. That's why I believe the "urban canyon" analogy is not quite good enough. It over simplifies the complexity of this problem.

Linear warfare: Urban fights will likely be linear on narrow fronts. However, we've got to be real careful. We don't want to be linear in our thinking. And there's a difference between the two in my view. You can be in a linear fight and not be linear in your thinking. The threat may design a plan to keep us in the urban area. They want us to get in, but don't want us to get out—the "roach motel" analogy. It's the old bait and ambush tactic that the Vietcong taught Lieutenant Ernst a long time ago.

We're not going to bypass cities. We'd like to. Everybody since Alexander the Great has said they want to bypass cities. Alexander the Great couldn't do it because his line of operations depended on those cities. Stability missions may cause you to go into the city because that's where the stability problem is. If the threat is your objective, then you're going to have to go where the threat is. So why don't we just stop talking about bypassing the cities and assume we're going to be in them? We're going to go there. Cities will always be there, that's an article of faith and a given. We need to be able to operate in them.

And we have to train at all levels. We have to be able to link them. We ought to be able to link a JTF at JFCOM to the ranger regiment to train for a mission using a simulation. We need to focus on the combined arms team commander, from brigade to joint task force level, because if they don't have their end right, we could dump a ton of money into the lower tactical level end of the problem and never succeed. This is what has got to happen. Whether we want to go in cities or not, we've got to be comfortable there. We've got to understand that it's an asymmetric problem. We've got to be comfortable there. Because when the generals and colonels don't get it right, we're not comfortable there, and the result is what happened on October 3–4, 1993. On our wall at the Joint Task Force Somalia headquarters, we had the name of every soldier, sailor, airman, and marine that was killed in Somalia. Every day those names reminded us of our responsibility—to get it right, to take care of soldiers. And that includes making sure they're ready to fight.

THE URBAN AREA DURING SUPPORT MISSIONS CASE STUDY: MOGADISHU

The Tactical Level I

SFC Matthew Eversmann, U.S. Army

The Tactical Level I

SFC Matthew P. Eversmann

Task Force Ranger

Bravo Company, 3/75 Ranger Regiment

Rangers Lead the Way!

I am SFC Matt Eversmann. In 1993 I was a squad leader in 2d Platoon, B Company, 3/75 and a member of Task Force Ranger. This was my first experience on a real world mission. This morning I am going to share my experiences in combat from a young leader's perspective. I was 26 at the time and had been in the Army for five years.

There is a lot to say and unfortunately my time is limited. This is not "the" story of the Battle of the Black Sea, but it does provide a little insight about what happens at the lowest level of troops. I will be happy to answer *any* question you have at any time during the presentation.

AGENDA

- **Arrival in Mogadishu**
- **Mission preparation**
- **Insertion on 3 October/Murphy**
- **Chalk 4 Successes**
- **Lessons learned**
- **Changes**

I have broken down my presentation into six areas. The first four slides will cover the four phases of the operation:

- Arrival in country
- Mission preparation at the hangar
- Initial insertion on 3 October 1993 (in particular, what went wrong for our chalk)
- What went very well.

The last two slides will cover my personal lessons learned and what I would pass on to the rest of the Army and sister services to help with the next urban combat operation. Please keep in mind that these are *my* thoughts. Everyone's impressions under fire can be different. What one person sees may differ from what his ranger buddy across the street sees. Also, I will add that some of my thoughts may differ from what you read in *Black Hawk Down* or on the Internet.

First Combat Experience
Arrival in Country

- **Information about flow is critical, ALL men must be informed within limits of OPSEC, etc.**
- **Must have a clearly defined task and purpose.**
- **Simple is BEST; it alleviates confusion.**
- **Focus on the TEAM for everything.**
- **Nineteen and 20-year-olds versus 30, 40, 50+ "decision makers."**
- **Leaders must be very aware that effects of every decision are "amplified." Be aware of the men; they WILL follow.**
- **Training stops at N-Hour.**

Information is key. You can never have too much when the stakes are real. This includes everything from hard intel updates concerning the operation to what is going on with adjacent units to what is going on at home. It is very reassuring to be "in the loop" as much as possible. I understand the need for OPSEC, but ANYTHING from higher to the men in a base area is crucial.

The bottom line for this mission is that we had a very clearly defined task and purpose: "Chalk 4 will occupy blocking position X. You will seal off the objective and not allow ANY interference on the target. Do not let anyone in or anyone out." Pretty simple. It gets hazy in a joint operation when there is dissension as to who is in charge of what and when there are questions regarding when authority changes hands, i.e., X is in charge until the infil[tration], then Y takes over on the objective, then Z takes command.

Chalk 4 understood its task and purpose very well. The chalk leader at the time, SFC Chris Hardy, and I started working our piece of the puzzle by breaking everything down to the simplest terms and actions. His foresight was key. No one knew how they would react under fire and it seemed that the less complicated a drill, the better. We basically rehearsed and rock drilled every piece of the operation from loading the aircraft, to infil[tration], to actions on the objective, to exfil[tration] and beyond. Those with previous combat experience and more specialized training were very good about sharing their knowledge with the rest of us. We were all part of the team. In relatively short order we were able to make and implement easy SOPs to accomplish our task. When doing this we included *everyone* on Chalk 4 in the planning. Everyone had a good idea; the good ones we kept; the bad ones we deep-sixed. We were able to make every event a *team* focus. Left side does this; right side does this.

I think it is key, and fortunate, that we were a close-knit organization before getting in country, that every man felt needed and vital to success. This may seem insignificant to an outsider, but you must realize that many on this task force had seen combat before. Those higher in the chain of command had a very distinguished combat history. What a commander or senior NCO thought to be a "nonevent" was, nonetheless, a significant event for a 19- or 26-year-old in combat for the first time. Because of this, every decision is "amplified" by the time it reaches a private. If the command is upset, the troops are very upset. If the command is dejected or sullen, the men are miserable. Remember, this was a whole host of firsts for most of us and anything perceived as negative or bad/harmful would not help the men's anxiety about being shot at. This having been said, we all understood the significance of this operation and everyone realized that "if we go tonight to get Aideed, then we go tonight to get Aideed. No more training!" We did conduct some outstanding training as time went on, but when we left Fort. Bragg, everyone realized that this was it. The alert starts; training stops.

First Combat Experience
Mission Prep

- **No real intelligence network.**
- **Technology versus the Third World.**
- **Soldiers are very astute when the mission is real.**
- **Warrior mentality.**
- **Refine the plan at every opportunity.**
- **Make money with rehearsals.**
- **EGS/contingency plan (as deep as possible).**
- **ROE must be clearly defined and understood.**

We found out that starting this operation was going to be difficult as there was no real intelligence network established at the time, at least as far as I was privy to. But I had no worries. After all, we were well armed, well equipped rangers and this enemy was living in the stone age. That was my thought on the Somalis. Consequently, I don't think that anyone had a "fear" of not getting back home. Certainly it was scary, but it was exhilarating at the same time. I think that I was more concerned about someone doing something wrong tactically than anything else. Please understand too, this is relative; we certainly were all concerned for the welfare and safety of the men. But if you could rate it, the thought that we wouldn't bring every ranger home alive was not in my mind. The men were consumed with doing the mission right and making it count. From disassembling their magazines daily to ensure the springs didn't rust, to retaping their fast rope gloves, to a host of other seemingly trivial details, the men were ever aware of what they needed to do and how important their individual mission was in helping the team and task force. This mentality is a good one. The warrior mentality should be fostered at every level. I think the task force commander did a very good job of cultivating that in all of us. After all, we all knew we were there to switch from safe to semi [-automatic] if need be. Going to war was all that we thought about and all we understood we were there to do. We weren't there to hand out bread or distribute fliers from the UN; we were here to get this bad guy and that was that. Having the force totally dedicated to that spirit was a key to success later on. As I said earlier, SFC Hardy and I continually worked on making our drills work more efficiently. With each operation we learned something new and unhesitatingly added it to our drills. FM 7-8, *The Infantry Platoon and Squad,* didn't cover what we were doing. The contingency plan at our level didn't go very deep. If someone was hit, then his ranger buddy had to take over. It was the same with the chain of command. We, the task force (TF), obviously had a combat search and rescue (CSAR) team dedicated to the operation. To me, that was good enough. Who would have thought that an aircraft would get shot down? Crash maybe, but not shot down.

A big concern was the ROE, again a very *real* first. It was very detailed and the situation was complicated by the fact that there were, for instance, some Somalis who were allowed to carry weapons (e.g., those working for the UN) and others that couldn't. Only the latter could be engaged. The former carried blue cards that gave them permission to carry a firearm, but from 100 meters a ranger could see only an armed Somali. It was a tough situation. The "what ifs" were running 100 mph. I realize now that ROE in the future may be very complicated. It took briefings from the staff judge advocate general (JAG, military lawyers) and the commander to spell out everything so that every 19-year-old understood the ROE. That is where the veterans were a big help. Everyone must look at ROE from the worst case scenario and backward plan from there. That makes the ROE easier to understand. What can I do when the bullets are flying and X happens? Do I shoot or not? It is that simple and that is what we all are paid to do.

3 October
Murphy with Chalk 4

- **Fall out one drill initiated.**
- **Bad part of town? Who knew?**
- **Wrong insertion point.**
- **Immediate Litter Urgent Casualty.**
- **FM radio destroyed/poor commo.**
- **No medic.**
- **Immediate enemy contact.**
- **Mission essential equipment.**
- **Mission accomplishment vs. casualty rate.**

3 October 1993 we launched on our mission. I became the chalk leader because our platoon sergeant had to leave on a Red Cross emergency. I assumed his role and went to the Joint Operations Center when the planning first started to unfold. Everyone said afterwards that "they knew that part of town was bad." Thanks for telling me. Maybe it was a blessing in disguise looking back. But that goes back to the information flow. I knew we loaded the rockets on the little birds (small special operations helicopters in support of the mission) this time and I saw the task force commander as we went to the Blackhawks. No cause for alarm for me.

Insertion. Everything changed for Chalk 4 after take-off. We were short of our location by at least 100 meters. We had to insert there due to the brownout conditions (due to the dust blown up by rotor wash), the resulting concern regarding collisions with other aircraft, and incoming enemy fire. The pilot told me and I acknowledged. It had happened before and we had a contingency worked for it. I was the last one out of the bird; by the time my feet touched the ground, my medic was working on PFC Blackburn. He had missed the rope and fallen almost 60 feet. He was in bad shape. On top of that, we were already taking fire from three directions. My radio telephone operator (RTO) had lost comms. His handset cable was destroyed and later we found a bullet in the radio (on the side that was against his back!). My alternate comms with higher was a hand-held Motorola and that was sketchy at best. No night observation devices (NODS) on any of us. Loaded for bear, but no NODS. One aid bag for my medic. I had taken out one of my canteens to secure more flash bangs (non-lethal grenades), demo, and flex cuffs, so I only had one quart of water. Because we were under fire so quickly, we couldn't "break contact" to get to our intended battle position. Nor could we get a MEDEVAC (medical evacuation helicopter) to our position. I hadn't planned on moving a casualty any distance for extraction. We had to stay and fight. We took three or four casualties by about three minutes into the fight. We went from thirteen rangers to eight by the time our convoy started to move to the crash site.

3 October
Chalk 4 Success

- **SOP and internal contingency planning.**
- **Medical training at Skill Level 1.**
- **Battle Drill / Muscle Memory / Marksmanship.**
- **Problem solvers.**
- **Little direction / Lots of initiative.**
- **Joint operation at shooter level.**
- **Find a way to win in spite of the enemy situation.**
- **Support up and down at chalk/squad level.**

Given what I have said about Murphy's Law and bad circumstances, I believe the men had tremendous success in dealing with adversity and persevering. We were able to execute the battle drill, even under fire for the first time, because we had made a simple plan that everyone knew intimately. Young rangers were able to shoot, move, and communicate under fire with little or no direction from me. I remember at one point of the battle doing a quick assessment of the perimeter and smiling as I watched the men aggressively engage the enemy from three directions in perfect standing operating procedure order. Our medical training was very good too. I had a private first class (PFC) emergency medical technician (EMT) opening airways, assessing damage, and basically, saving a ranger buddy's life under fire. I saw rangers dealing immediately with terrible wounds inflicted on their ranger buddy. It was instinctive. It was absolutely incredible. In this dire circumstance that every infantryman wonders about, the men reacted in textbook fashion. We had had a debate with my platoon leader about the S.L.A. Marshall claim concerning how many men would actually pull the trigger under fire. We had 12 of 13 and I have to believe that if he could have, Ranger Blackburn would have done the same thing. The men knew their sectors of fire; they knew the ROE, and they knew their task. We did not allow any enemy to pass our position. The men held off the enemy. (We later learned that a considerable number of Somali forces were attacking from the north.) We were the northernmost battle position (BP). We had help from the assault force (the group tasked to seize the Somali targets) in dealing with our casualty. No one questioned who was who, or what needed to be done. We all just executed. It can be done. I also saw some heroic actions by young men in combat for the first time. As I was trying to keep my cool and get our casualty extracted, seeing the men perform gave me the confidence and reassurance that I needed. The leader has nowhere to look but down, and if the men (well trained and disciplined) do the right thing, I *had* to be strong and vice versa. It was definitely a two-way street.

Lessons Learned

- **Good, well-trained men will die in combat. NOTHING REPLICATES THIS.**
- **"Team" must be solid BEFORE deploying.**
- **Inside is better than outside.**
- **Live fire exercises build cohesion and confidence. This took away the initial fright of enemy fire.**
- **Warrior mentality is CRITICAL!! TRAIN FOR WAR, NOT NTC, JRTC, etc.**
- **Everything must be related to combat.**
- **All soldiers must be intimate with urban operations.**
- **The 50 meter battlefield: It is 3D and fast.**

This is a brief synopsis of what I took away from Somalia. Sergeant Joyce died doing the right thing, as did the other TF members. That is a fact of warfare. Even if the enemy does live in the nineteenth century, sheer numbers will take over. Eight thousand enemy surrounded 150 men; it's not hard to understand this. And the way our adversary fought was what a westerner would likely consider unorthodox. For example, the Somalis would set up ambushes by placing half their force on one side of the street and half on the other; they would then face toward the middle of the street and shoot. We all have to understand that men dying is a fact of life in this profession. We couldn't have been successful if we had a dysfunctional team before deploying, not with such a small force. The ability to do live fire exercises is key. I think that most would agree that the confidence each ranger had under fire was proportional to the amount of live firing he had done. Shooting in close proximity builds confidence and cohesion as does marksmanship training. Being used to live firing and weapons handling in all types of situations was invaluable. If there is one thing to train on, or one tool to use, it has to be live fire.

Every soldier has to be committed to being a warrior or someone will die needlessly. Private to colonel, everyone must have their game face. When things go bad, if you don't have the trust and confidence in your subordinates and leaders, someone will die. *Everything* our military does should be geared toward war. We have to train as we will fight. Simple and redundant. Do not get caught up in the NTC-JRTC fight. Understand that some training conditions exist that preclude "real" scenarios. But the attitude of every leader and soldier must be going to *war*.

I also learned that it is much better to be inside buildings than outside in the city. It isn't always possible, but whenever the force can have more protection it has to be done as long as you can still accomplish the task. The 50 meter battlefield is very fast, very close, and very frantic.

Everyone needs to be able to assess the situation, and doing that while in the street with enemy fire incoming from several directions is extremely difficult. Being inside a building gives leaders and men a moment to evaluate a situation and think; the streets provide no such luxury. In the streets the fight is close; it is instantaneous; there is nothing harder in combat. Everyone at the staging area is potentially an infantryman: cooks, mechanics, MPs, anyone who is there when bullets fly must be thinking about *war*. Who would have thought that two Blackhawks would have been shot down in the city? God bless our cooks who stepped up and came to our assistance in the relief convoys. They literally put down the spoons and spatulas and grabbed their weapons to go to the fight. That is the warrior mentality. When we go again, it just might be that low-density military occupational specialty (MOS) soldier who becomes the last line of defense. He *has* to be prepared.

CHANGES

- **Joint Training whenever possible: SOF, mechanized, conventional.**
- **Sidearms for all soldiers.**
- **RANGER CREED APPLIES TO EVERY WARRIOR.**

If I could change some big picture things in the military this is what I would do. I would cross-train whenever I could with whomever I could. SOF with mechanized and SOF with conventional. The battlefield isn't the time to learn how to operate the hatch on an APC or how to use the Coax guns, let alone how to drive the vehicle. I believe we do this better. Not to step into the RSMs box here, but the Ranger Regiment has been, I believe, a great steward in helping our military by doing its fair share of cross training with other units.

I am a believer that every soldier should have a sidearm. This is a big issue, I know, but being in the back of a HUMVEE with enemy surrounding you and having to pass a 9mm around to whoever has a shot because the unit is short on ammunition is a bad, bad situation to be in. You can only plan so deep, and you can only carry a finite amount of ammo. I shot 13 magazines, as did most. In a situation where you must shoot or die, it is a terrible feeling to know that you are out of ammo.

Lastly, I believe that the Ranger Creed applies to every soldier in every service. The pride that it instills, the absolute certainty that it provides, is something that every soldier needs to be able to fight and win on any battlefield. I would encourage everyone to look at it, dissect it, and come away with nothing less than the warrior mentality.

God bless you all—Rangers Lead the Way.

Appendix N

THE URBAN AREA DURING SUPPORT MISSIONS
CASE STUDY: MOGADISHU
The Tactical Level II: The Offensive and Defensive Use of Urban Snipers

MAJ Scott D. Campbell, U.S. Marine Corps

Five Fleet Antiterrorism Security Team (Fast) platoons ultimately deployed to Somalia. Ours, 5th Platoon, was the second of the five sent to support the U.S. diplomatic mission in that nation.

Fast Platoon Organization

- HQ: Plt Cmdr-Capt
  - Plt Sgt-SSgt
  - Plt Guide-Sgt
  - Plt RTO-Cpl
  - Corpsman-HM2
- Sqds-3 x 13
- Sniper Tms-4 x 2
- Total Plt = 52

The platoon's organization is based on a basic marine rifle platoon. This platoon has enhanced skills and weaponry. It can be task organized and equipped to meet specific mission requirements.

This photo was taken in the U.S ambassador's residence in Mogadishu. Snipers employed bed sheets to allow them to blend in with the surrounding walls.

Sniper Weapons and Equipment

- 4 x M49 Spotting Scopes
- 4 x M14 DM Weapons (7.62 mm)
- 4 x M40 Sniper Rifles (7.62 mm)
- 4 x M16 H-Bar Rifles (5.56 mm)
- 4 x M16A2 Rifles (5.56 mm)
- 8 x M9 Pistols (9 mm)
- 4 x AN/PVS-4 Night Vision Scopes
- 4 x SIMRADs Night Vision Scopes

The sniper section had a wide variety of weapons available for employment. The M-40s and M-14s worked very well for daylight operations but the SIMRAD scopes were problematic at night. Humidity caused the scopes to fog up on a regular basis and limited the sniper's ability to engage targets at night. The H-BAR M-16s with ANPVS-4 scopes proved more capable. Permanently mounting the scopes on the H-BARs gave us a reliable night sniping capability.

This was a sniper position on the roof of the U.S. ambassador's residence.

Sniper Training

- Phase I : Individual training
  - Sniper School, Quantico, Va
  - Division Sniper School, CLNC
- Phase II : Section training
  - Scenario training
  - Sustainment training
- Phase III : Platoon integrated training
  - Security Operations/Escort Operations
  - Recovery Operations

Snipers conducted their training in phases. Upon completion of the basic schooling the section began scenario-based training. This training consisted of both urban and rural operations and included defensive as well as offensive conditions. Scenario-based training utilized shoot/do not shoot situations. Sustainment training consisted of firing on known distance ranges up to 700 yards at least twice a month. During platoon integrated training, a considerable amount of time was spent on recovery/target-specific training in support of an assault element seizing an objective. The sniper section was also trained in sketching, range card development, and the use of supporting arms.

Mission

- Conduct compound and mobile security to defend State Department and other associated personnel IOT provide a safe environment for the conduct of diplomatic operations.

This was our mission statement as I remember it. Our deployment order originated at the JCS and assigned us to CINCCENT. We were under the operational control of the U.S. diplomatic mission. Our day-to-day orders were driven by the ambassador or the senior Diplomatic Security Officer.

Tasks (Specified)

- Provide security for both the embassy and housing compounds.
- Provide mobile security for helicopter and motorcade movements.
- Conduct other tasks as directed by the ambassador or the head of the diplomatic security detachment.

The tasks were in keeping with our training and capabilities. The greatest limitation to providing proper security was our inability to patrol outside of our defensive positions. This prohibition, combined with no indirect fire capability, limited our ability to counter potential threats. Although the lack of indirect fire assets was a problem, it did make sense in light of the urban environment and ROE. Helicopters were made available during escort missions for use as transportation and emergency fire support.

Tasks (Implied)

- Provide support to United Nations forces within the scope of capabilities and area of influence.
- Deter hostile action.

Note: No offensive taskers

There were no offensive taskers. The majority of our engagements were in support of U.S. or UN forces.

This sniper hide was constructed using white mesh. The mesh was very effective in hiding teams from observation. The teams were virtually invisible from the street below. The mesh in no way hindered the teams ability to observe or engage hostile targets.

Rules of Engagement

- Required positive identification of militiamen/clan members in the act of committing aggressive/hostile action IOT allow for friendly/UN forces to engage/apply deadly force.

The ROE were not a major problem. The biggest problem was how the marines applied the ROE and what they understood them to mean. Aggressive or hostile action can be different things to different men. Initiating an engagement initially required my or the platoon sergeant's authorization. As the marines became more familiar with the ROE and how they were to be applied, we were able to allow them to engage on their own initiative. At no time was the validity of any engagement questioned by the UN or the diplomatic staff. The snipers exercised considerable discipline in the use of force and did not engage on several occasions when they could have.

This photo shows two sniper teams developing sketches and range cards.

Sniper Operations (Phases)

- July-August: Reactive, deployed teams in response to hostile action.
  - Developed detailed range cards, identified sectors and "coded" associated buildings.
  - Tracked movement of personnel in our area of influence and identified dead space.
- Sept-Oct: Proactive, aggressively employed snipers to deter and eliminate hostile action.

During the early stages of the deployment the snipers were reactive. We underwent a learning phase during which the marines became more comfortable with the ROE and began developing ideas on how they could be employed more effectively. Initially, hostile action was limited. This period allowed the men to familiarize themselves with their environment and construct their positions. As hostilities escalated, the teams began to understand the cycle on which the Somali gunmen worked. This cycle revolved around sleep, drug use, and certain periods during the day when the gunmen were more active. As the situation became more hostile, the snipers began engaging with more frequency and therefore were a major influence in controlling the level of hostilities on our section of the perimeter.

Method of Employment

- Independent/section sniper operations
- In support of convoy operations
- In support of general engagements
- Limitation: Not authorized to employ snipers outside the UN/embassy compound.

The primary function of the snipers was to aid in the protection of the compounds; much of their work to this end was independent of the daily guard routine. Including the snipers in convoy operations was essential. Although their value was limited during movements, they were invaluable in supporting security operations once the convoy arrived at their destination. Because we routinely visited the same locations, we were able to develop detailed sketches of most sites and determine the best places to employ available assets. The snipers were not employed in hides when supporting general engagements. We learned early that firing more than one shot from a hide compromised that position and generally drew enemy fire. By employing them from positions with other marines we were able to engage more than once if the situation required. Employing sniper positions outside the compound was not an option due to our mission. The nature of the environment was not conducive to clandestine sniper operations.

Independent/Section Sniper
Operations

Independent/Section Sniper Operations

- Issued section sniper order that was modified as the situation changed.
- Conducted sniper operations within sectors to deter and eliminate hostile acts.
- Coordinated with UN forces to employ snipers within their sectors.
- Identified and prepared hides.

The initial order that was given to the sniper section required frequent modification as the situation changed. We modified our operations as we refined methods of employment. Because of the fixed nature of our positions, we had a limited number of areas that were suitable for the construction of hides. Periodically we overtly employed snipers to act as a deterrent. In addition, we constructed dummy positions and moved personnel to give the appearance of a more robust sniper capability. Coordination with UN forces inside the old embassy allowed our teams to occupy positions that supported the embassy and housing compounds but did not expose the teams to threats outside the main UN perimeter.

Convoy Operations

- Location within convoy
- Site review/plan for employment at final destination
- Assignment of sectors of fire/observation

As discussed earlier, at least one sniper team accompanied all convoys. The snipers generally occupied a seat in the last vehicle in the convoy. Based on the nature of the threat and our experiences with being ambushed, we learned that the Somalis generally engaged the lead vehicle first. Having the snipers at the back of the convoy would often give them the opportunity to dismount and engage. In general, however, our convoy SOP was to not stop the convoy if ambushed but to drive through and out of the kill zone.

General Engagements

General Engagements

- Not employed in hides
- No limitations on number of engagements
- Employed to cover avenues of ingress/egress and isolate or limit flow of forces in our area of influence
- Attempted to add depth to engagement area

An additional reason for employing the snipers with the other marines during general engagements was we were able to position the teams in areas that allowed them to observe and engage targets at greater ranges. The majority of the sniper engagements were at a distance of 200-300 meters. By positioning teams so that they could see down avenues of approach, we could better assess the flow of hostile forces into our area of influence and add depth to the engagement area. We saw that employing them in this fashion, although overt, allowed the teams to act as combat multipliers much like artillery and mortars by limiting the enemy's ability to move men around the engagement area.

Basic Rules of Employment

- One shot engagements (protect limited hides)
- No discussion of engagements on radio
- Immediate debrief/critique
- Two hour time limit in hides
- STAY WITHIN ROE. IF IN DOUBT, DO NOT SHOOT.

Problem Areas

- Night Optics
  - SIMRAD vs AN/PVS-4
- Intelligence
  - Intelligence geared towards the capture of Aideed, not towards compound security or convoy operations
- ROE
  - Countering hostile reconnaissance effort
  - Enemy use of civilian populace.

Difficulties with night optics have been discussed. Intelligence was not geared toward convoy operations or base defense. Although many requests were submitted for information support, the intel community was focused on gathering information that would aid in capturing Aideed. Basic information concerning times and locations of enemy activity were not forthcoming. We developed our own event matrix that depicted the times of day that had more significant enemy activity. In addition, we were able to map the locations of all incidents, determine what areas were most prone to enemy activity, and identify the nature of that activity. This, combined with our ongoing observations of local activity, allowed us to employ our snipers during peak periods and become more effective. The biggest problem that we had with the ROE was our inability to engage unarmed men that were obviously conducting reconnaissance of our activity. Somali men would get on adjacent roof tops and overtly observe our positions. If armed, the situation might mature and allow for an engagement, but most often they were unarmed and we were powerless to prevent this activity. In addition, during protests/riots near our positions, armed Somalis routinely mixed with the crowd and proved to be difficult targets to engage.

Lessons Learned

- More than one shot from a hide generally compromises the hide and draws fire.
- Scenario based training vital for applying ROE.
- ROE interpretation
- Mesh screening invaluable for hide construction in buildings.
- DO NOT MEASURE EFFECTIVENESS WITH BODY COUNT.

Many lessons were learned from our experiences in Somalia. The most important lesson I learned was that we should not measure effectiveness with a body count. Snipers are aggressive, goal-oriented individuals. By counting kills you put the focus on the wrong goal and risk creating competition. In a restrictive urban environment we must avoid creating situations that can obscure the objective of the operation. My solution was counting all engagements as sniper section events; it was always a team effort. Although I am sure the sniper kept count, we never discussed engagements in individual terms.

Appendix O

THE URBAN AREA DURING SUPPORT MISSIONS CASE STUDY: MOGADISHU

Medical Support

LTC John Holcomb, U.S. Army

Urban Trauma, Lessons Relearned In Somalia, 1993

LTC John Holcomb, MC
Director, Joint Trauma Training Center
Ben Taub General Hospital
Houston, TX

Proud and Honored

Panama 1989

Somalia 1993

Green Ramp 1994

I am proud and honored to have taken care of soldiers injured in combat. These experiences include Panama in 1989, Somalia in 1993 and, while not a combat action, the injured during a mass casualty situation at Ft. Bragg, NC in 1994, what we call "green ramp." It's an amazing experience and one that is not replicated in any civilian trauma environment. Unfortunately, in a sense, you do it better the more times you experience it.

Overview

- Lessons learned / relearned / rehashed over the last 6 years
- Medics ➡ Surgeons
- Medical Hotwash 2–3 days after battle
- SOMA Conference Dec 1998
  - April 00, *Mil Med* review article
- CPT Bob Mabry, MC
  - SFC Bob Mabry, 18D
  - *Journal of Trauma* review article

In the next 45 minutes or so I will go over lessons learned and relearned from the last six years. These lessons come from speaking with numerous medics, nurses, PAs, and surgeons. A very important part of this overview was a medical hotwash that occurred after the October 1993 battle in Somalia. We included every medical care provider from the youngest private to the most senior surgeon. Together we went over every casualty from the point of injury back to the casualty collection point to the hospital, including the evacuation to the airfield where casualties boarded C-141 aircraft and flew to Germany. Let me say at this point that the medics did a superb job and in every situation treated every case exactly as they should have. Further reviews of these cases and scenarios have gone on over the past several years, culminating most recently in the December 1998 Special Operations Medical Association conference during which we reviewed multiple scenarios. The results were published in the April issue of *Military Medicine* and a supplemental review article.[1] Captain Bob Mabry, who in 1993 was Sergeant First Class Bob Mabry, an 18 Delta working with one of the special units that was deployed to Somalia, has recently published his *Journal of Trauma Review* article dealing with a number of the same issues.[2] This is a significant event; this is a very difficult review article to publish because there are no prior cohesive, unified lessons learned from a medical deployment contingency.

---

[1]F. K. Butler, J. H. Hagmann, et al., "Tactical Management of Urban Warfare Casualties in Special Operations," *Military Medicine,* Vol 165 (2000) pp. 1–48.

[2]R. L. Mabry, J. B. Holcomb, A. Baker, J. Uhorchak, C. Cloonan, A. J. Canfield, D. Perkins, and J. Hagmann, "U.S. Army Rangers in Somalia: An Analysis of Combat Casualties on an Urban Battlefield," *Journal of Trauma,* Vol 49 (2000), pp. 515–529.

On 3 Oct 1993, during a daytime raid, U.S. troops were caught in a firefight; helicopters were shot down; a resupply column was ambushed; and eventually casualties started arriving at the 46th CSH.

On October 3, 1993, U.S. troops were caught in a firefight (during a daytime raid). Helicopters were shot down; vehicles were ambushed, and eventually casualties started arriving at the American combat support hospital. We've heard many of the details and descriptions of these events; discussions regarding the tactical situations are better left to those who were actually on the ground. I will speak from my point of view at the 46th Combat Support Hospital where I was deployed for about 30 days as a general surgeon with the Ranger Task Force but assigned to the hospital.

46th Combat Support Hospital

- 200 combatants
- 112 injured
- 70 casualties hospitalized
- 42-bed hospital (100 personnel)
- 2 waves
  - 1730 on the 3rd
  - 0600 on the 4th
- They did an outstanding job.

The 46th Combat Support Hospital handled around 200 combatants, 112 injured casualties. Seventy casualties were hospitalized. The facility was a 42-bed hospital with a staff of approximately 100 personnel. We had two waves of casualties, one starting at 1730 on the afternoon of October 3rd and the other at 0600 on the 4th. Again, let me say that the 46th Combat Support Hospital did an absolutely outstanding job and unfortunately was never officially recognized for the outstanding care they provided to these special operations forces. They took care of more casualties in a shorter period of time than any U.S. deployed military medical asset since Vietnam.

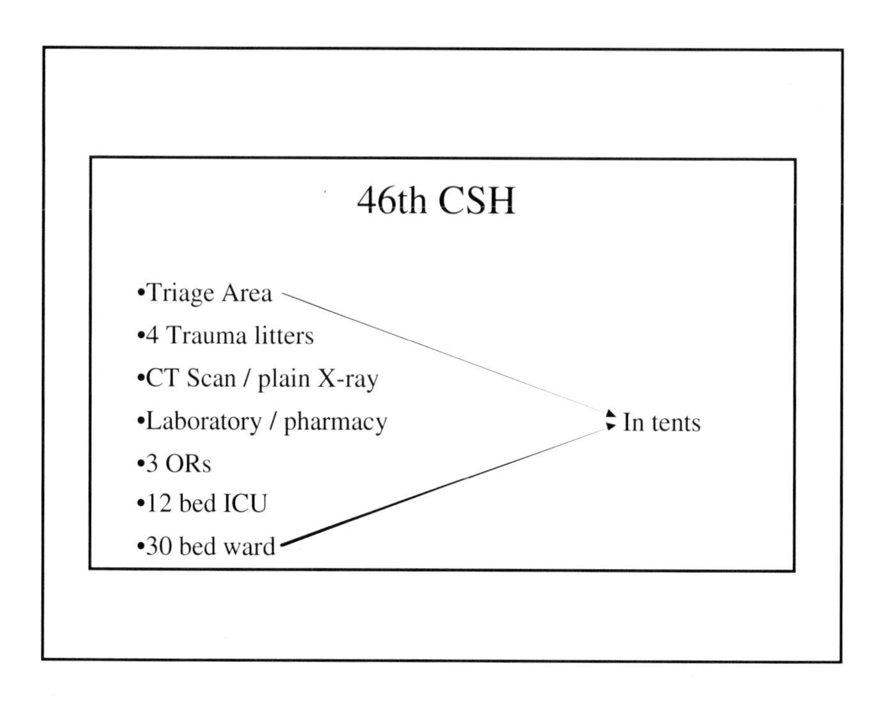

The 46th consisted of a triage area at which they had 4 trauma litters; a CT scan, plain x-ray, laboratory and pharmacy, three operating rooms, a 12-bed IC unit, and a 30-bed ward, all in tents. All in all a very capable hospital.

Aerial photo of the 46th Combat Support Hospital setup on a corner of the UN compound.

3–4 Oct Mass Casualty Event

34 Cases — 36 Hours — 3 Surgeons

7- Laparotomies (1 neg)	(20%)*
1- Sternotomy	(3%)
2- Neck dissections (1 negative)	(6%)
2- Burns (1 inhalation injury)	(6%)
2- Completion Amputations	(6%)
9- Soft tissue washouts	(26%)
13- Open fractures	(38%)
112- Total Casualties	(Many other providers)

* Percentages do not add to 100 due to rounding.

On October 3–4, 1993 mass casualty event, 34 operative cases were performed over 36 hours by 3 surgeons. You'll notice that the distribution of injuries runs the gamut from laparotomies, or abdominal exploration, to chest operations and neck operations; there were burns; there were amputations and open fractures. The percentages on the right hand side have a distribution very similar to any that you see in textbooks regarding war injuries. There were 112 total casualties. There were many other providers that assisted with the non-operative and operative cases.

Lesson

- Deployed general surgeons need to have broad training and be allowed to practice elective vascular and thoracic surgery at their local hospitals—so that they will be ready for those cases (20% of all cases) in combat.
- Read all medical personnel

The lesson learned from this distribution of casualties is that deployed general surgeons need to have broad training and be allowed to perform elective vascular and thoracic surgeries at their local hospitals so they will be ready for those cases, approximately 20 percent of what they will see in combat. This is critically important. At our MTFs during peacetime the general surgeons are the most forward deployed surgeons. They are generally the youngest surgeons who are not allowed to perform elective thoracic and vascular surgery because of local peacetime civilian constraints due to scope of practice, training, and credentialing issues. However, when these same surgeons are deployed they have to take care of these very significant injuries under the most austere conditions, never having had the ability to perform those same operations back home in an elective situation. That is not correct. That's not right. That's not fair to surgeons and not fair to casualties. This really devolves down to sustainment of skills and sustainment of skills is a very critical problem for all deployed medical personnel, be they medics, nurses, or surgeons.

Historical Causes of Death on Battlefield
(KIAs = 20%)

- Head Injury
- Hemorrhage
- Tension pneumothorax

The preeminent historical causes of battlefield deaths are head injuries, hemorrhage, and tension pneumothorax. This yields a KIA rate of about 20 percent, a rate that has not changed over the last 150 years.

Causes of Death in the Hospital
DOWs = 2%

- Sepsis
- Head injury
- Hemorrhage

The primary causes of death in the hospital (died-of-wounds) are sepsis, head injury, and hemorrhage. That yields a died-of-wounds rate of two percent. That statistic has remained essentially unchanged since the end of WWII.

Somalia October 3, 1993
Killed in Action
12%

Cause — Unknown
No autopsy data
Body armor?

*Small Numbers

In Somalia in 1993 our killed in action rate was around 12 percent. The causes are unknown because there were no autopsies done. There were only surveys. One of the reasons that the killed in action may have decreased from 20 to 12 percent is that there was effective body armor in use. However, these are very small numbers and it's hard to make broad generalizations from such a small data set.

Somalia October 3, 1993
Died of Wounds
3%

Pelvic Wounds (2)
Chest / Abdominal injury (1)
Trans-abdominal (1)

*Small Numbers

The died-of-wounds rate was three percent from pelvic wounds, chest and bowel injury, and trans-abdominal wounds. These, again, are very small numbers from which it is hard to make a broad generalization.

Lessons

- KIA may be lower because of effective chest body armor
- We have not had a great impact on DOWs in a long time
  - but the number is small
- Direct military research monies to decrease KIA rate

As noted, the KIA rate in Somalia may have been lower because of truly effective body armor. We have not had a great impact on died-of-wounds in a long time; we need to direct military research monies to decrease the KIA rate, which has remained a very high 20 percent for 150 years.

This casualty's body armor clearly saved his life.

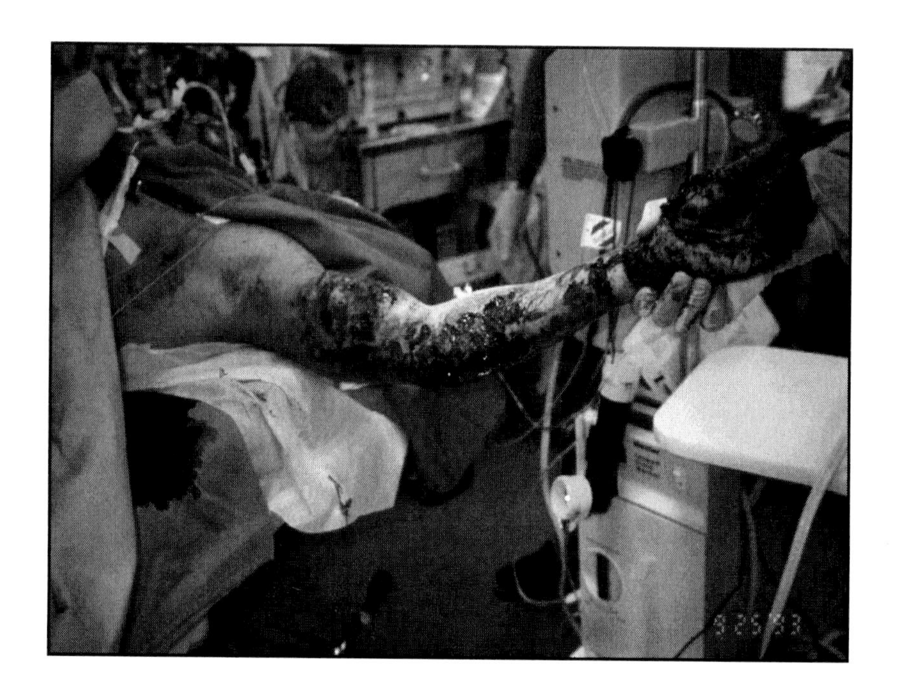

A casualty who had an anti-tank weapon go off in his right hand.

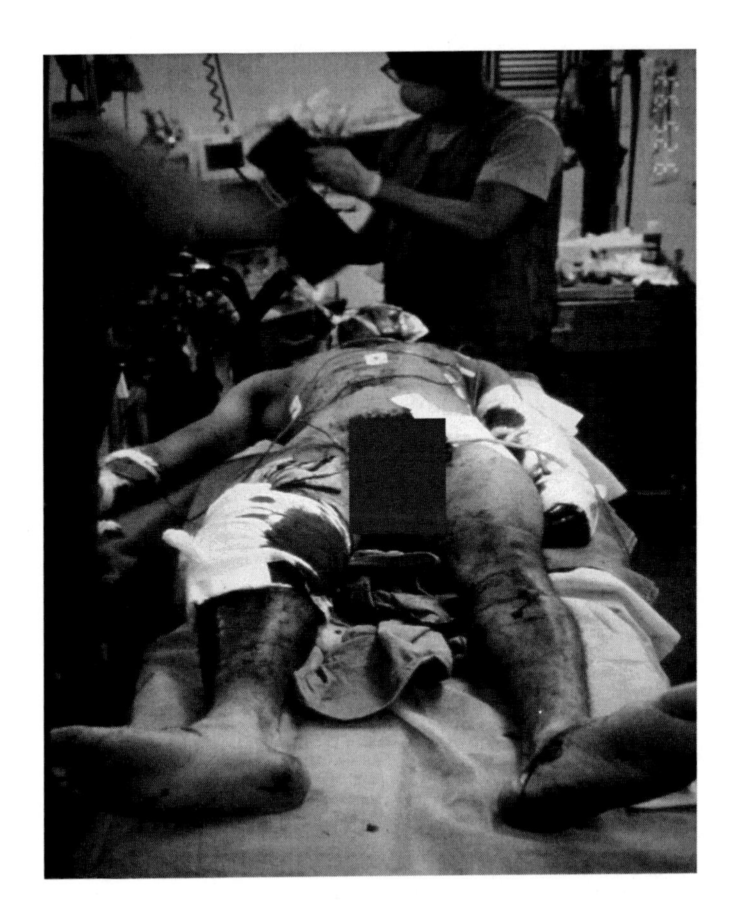

It is notable that there are no injuries to this soldier's chest or abdomen where his body armor was. These slides show examples of extremity injuries with no major trunk or abdominal wounds. It is another example of body armor saving a man's life.

What's the Medical Problem?

Expectation of State of the Art Care in the Middle of Nowhere

Non-permissive
Triage
Experience
Equipment
Personnel
Numbers
Logistics

What's the medical problem? The expectation is state-of-the-art care in the middle of nowhere. However, there are a lot of problems with state-of-the-art care in the middle of nowhere. It may be that the care must be provided in a non-permissive environment. You may have to perform medical triage, medical triage with large numbers of casualties and minimal people. Those people may not be experienced with combat casualties. Your equipment is different than what you use to practice medicine in a peacetime facility back in the States. The number of personnel that you have may be inadequate to take care of the number of casualties in a traditional sense. And the logistics of your supply system may be inadequate to support the casualties you have coming in. All in all anyone would agree that taking care of casualties out in the field is very different than taking care of casualties in a civilian hospital.

Lesson

- The FDA and JCAHO do not exist in the field.
- That doesn’t mean we don’t try to provide stateside level care to our casualties , but some reality must intervene.

The lesson here is that the FDA and JCAHO do not exist in the field. These regulatory bodies and the regulatory environment they create in a civilian hospital do not apply in the field. It doesn’t mean we don't try to provide a stateside, state-of-the-art level of care to our casualties, but at some point our leaders must recognize that reality has to set in, that we can not attain that level in tents in Africa, Korea, Eastern Europe, or wherever we may be.

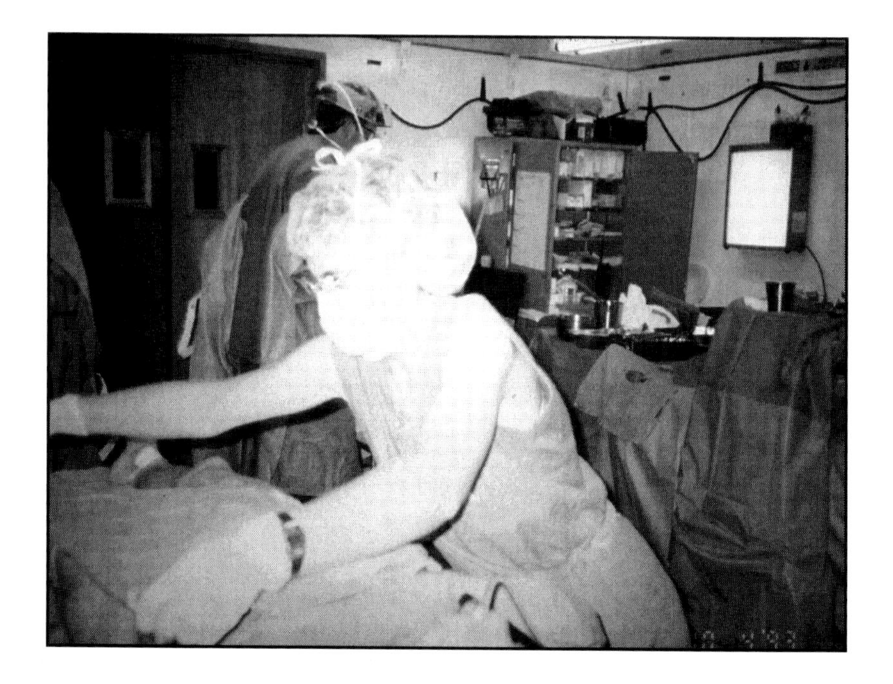

This is the reality of care in the field. It's a very nice environment. You can take care of casualties very well, but you have one nurse for two OR tables that are four feet apart. This does not happen in the United States. It is important that leaders stand up and tell the FDA and JCAHO that the rules do not apply in the field.

Types of Injuries / Problems

Shrapnel

High Velocity

RPG

Mines and Bombs

Delayed Presentation

Contaminated

Overwhelming Numbers

The types of injuries and problems that you see in the field include those created by shrapnel, high-velocity rounds, RPGs, mines, and bombs. These types of problems fortunately do not apply to civilian trauma. We are not seeing RPGs, mines, and bombs in the civilian world yet. Occasionally you see a high-velocity weapon, but not very often. Certainly we rarely see overwhelming numbers of severely injured casualties. This is very unusual in the civilian world.

Lesson

- These injuries are different from civilian casualties and the military must teach their medical officers and NCOs the lessons learned from past conflicts.
- There is no coherent medical lessons learned center for this type of information.

The injuries are very different from civilian casualties. The military must teach the medical officers and NCOs the lessons learned from past conflicts. Unfortunately there is no coherent, well-organized medical lessons learned center for this type of information. That is a travesty. It's a tragedy that needs to be fixed.

Medical Capability in the Field

What do you really need?

What do you really have?

What can you really do?

Medical capability in the field. Those questions are, what do you really need? What do you really have and what can you really do? For the most part, the more experience you have with taking care of trauma patients, the more you realize that you actually need very little equipment. You can do a lot with very little.

Lesson

- Especially for those with no prior field experience
  - a lot of good can be done with a little bit of gear
    - CT?
    - That special instrument?
    - That special person?

This is a hard lesson learned, especially for those with no prior field experience. The solution is to improve the trauma training and the trauma experience in the military. Our central core is taking care of trauma patients, not retirees with cancer. Not dependents with hernias. Not breast cancer in our wives. Taking care of trauma patients is the central core of military medicine. We need to realize that and we need to start training for those eventualities.

MOUT Casualty Numbers

- 50% Casualties
- We rarely train for these numbers
  - leaders are not prepared
  - tactical implications
  - medical system not prepared
    - immediate care
    - evacuation
    - resupply

Source: COL Cliff Cloonan

Casualties during military operations in urban terrain run at the 50 percent rate. We've heard from the operators at this conference how devastating this is. However, I can tell you that in nine years of special operations deployments not once did we ever see a 50 percent casualty rate exercised in an exercise. The leaders are not prepared. The tactical implications are not addressed and the medical system is not prepared to receive this number of casualties. However, any time you read about urban terrain casualties that 50 percent number keeps popping up.

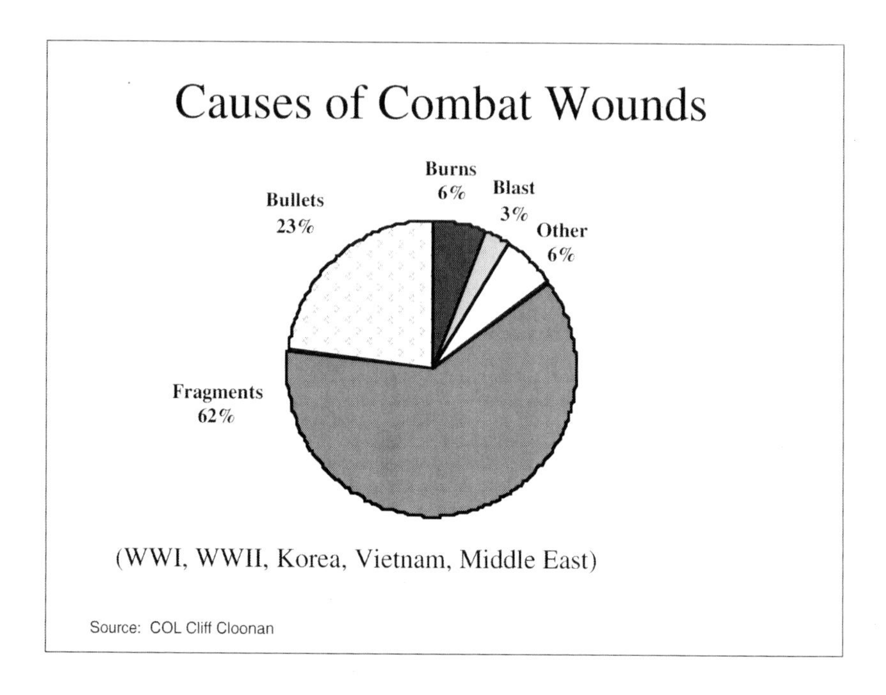

The cause of combat wounds for over 50 percent of the casualties is fragment wounds. Bullets count for less than a quarter, with burns, blasts, and other causes comprising the remainder. This has been demonstrated from WWI through and including combat in the Middle East in 1991.

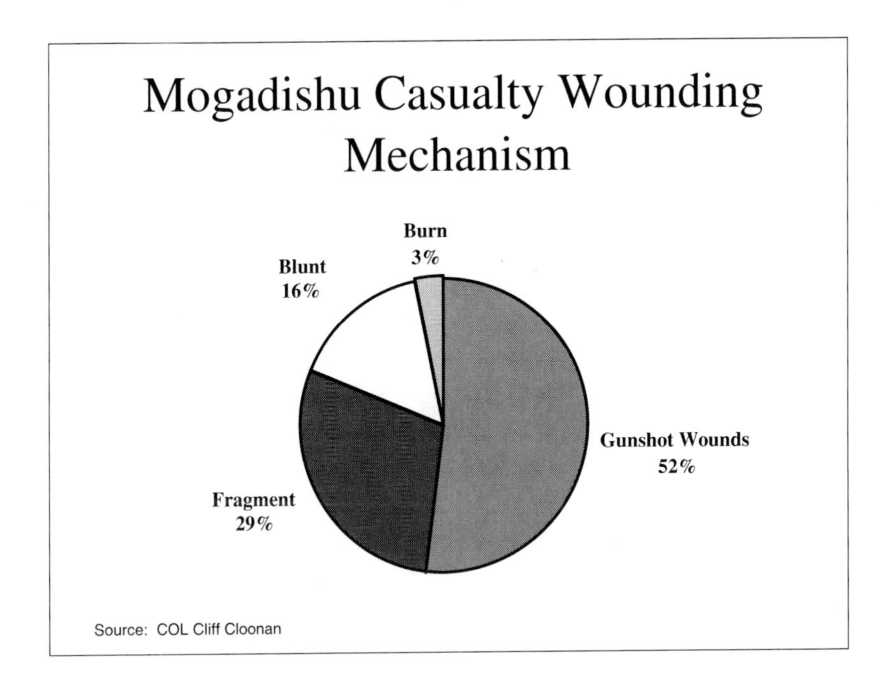

In urban terrain, gunshot wounds become more predominant with over 50 percent of the wounds being from isolated gunshot wounds, fragments dropping to 20 percent. And blunt injuries are not an insignificant percentage. Burns are a very small percentage. Why is this important? It makes a difference for how we take care of the patients. Single gunshot wounds are handled very differently than multiple fragments.

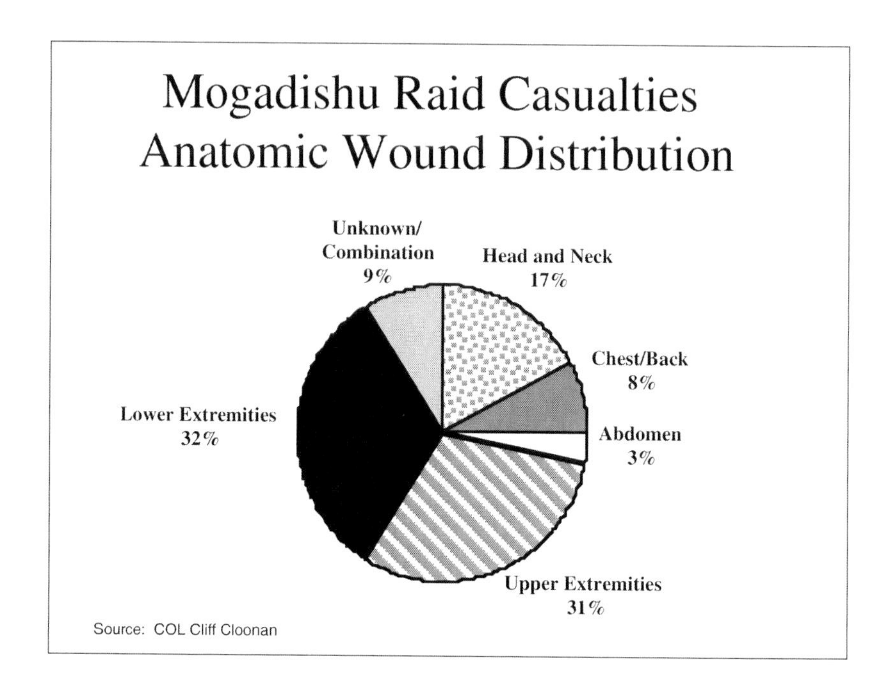

Along with these problems, with these different wounding mechanisms, are evacuation delays in which no evacuation may be possible for a long time.

USSOCOM Sponsored
Conference on Medical Lessons
Learned Mogadishu
Dec 1998

Special Operations Medical Association
CAPT Frank Butler, MD, USN
April 00 Supplement in Mil Med

Reference: Butler FK, Hagmann JH, et al. Tactical Management of Urban Warfare Casualties in Special Operations. *Mil Med*. 2000; 165 (4, supp: 1–48).

What Makes Medical Care In This Environment Different?

- Single GSW
- Wound contamination
- No evacuation possible or
- Long evacuation delays
- Hostile care environment—Fire superiorit y is best medical care in a firefight
  - "Reverse Triage"—Treat least wounded first to maintain firepower

As the shooters know, in a hostile care environment fire superiority is the best medical care. Reverse triage, where you treat the least wounded first to maintain the firepower, is currently being employed in some operational units.

MOUT Lessons

- Crowds can overwhelm
  - fixed wing (AC-130) could help
- Medical recommendations have tactical impact
  - call for evacuation

In this environment, a medical recommendation to call for evacuation has very, very significant impact. Leaders at small unit levels need to recognize that evacuation recommendations come from their medics. One of the questions that needs to be answered is what is the effect of delay on the casualty? If we can answer these types of questions we can give the tactical leaders better information. In other words, we can recommend evacuations but maybe delay an evacuation as long as it's not going to hurt the casualty. That tactical leader may have to decide that hurting that casualty by delaying evacuation is the operationally correct decision. These types of decisions need to be exercised in the light of day and discussed and implemented on exercises so that these young leaders do not have to consider the decision for the first time at 2:00 a.m. during a firefight.

Evacuation

- Move casualties to a "safe" spot
- Extraction of casualties
  - don't generate more casualties
- Delayed evacuation
  - up to 15 hours (so train for it)
  - effect of dehydration
- Medical resupply
  - type, size, number of bags of fluid
  - all others

Evacuation consists of moving casualties to a safe spot and doing so without generating more casualties. Delayed evacuations of up to 15 hours is a very common theme when you read about MOUT casualty care. So we have to train for it. There is a significant danger of dehydration. Medical resupply in these environments can be very, very difficult. Running out of bags of fluid, bandages, and other typical medic-carried supplies is a real possibility.

Evacuation

- Helo evacuation and ground = very difficult
  - RPGs, LZs, ambush
  - ability to overcome roadblocks
  - armored vehicle?
  - ability to maneuver
  - safe transload when vehicle disabl ed

In Mogadishu, helicopter evacuation and ground evacuation were both very difficult. RPGs ambushed these two types of vehicles very easily—as we've heard from the operators. A road block of tires essentially stopped the ground evacuation, and we know of the problem regarding helicopters getting shot down.

Random Issues

- Armored floor mats
  - excellent cover for CSAR team
- Non-lethal vs. effective application of conventional weapons
- Pain medicine that does not disable
- Should obviously dead be left behind or recovered?
- Impaled unexploded ordnance
  - *Military Medicine*, 1999

An issue that came out of the Special Operations Medical Association Conference: armored floormats make excellent cover for a CSAR team. They can be torn out of downed helicopters. Use of non-lethal weapons in lieu of lethal munitions was not very popular at the Special Operations Conference. There was a discussion of pain medicine that did not disable but allowed the combatants to continue to safely carry their weapons. Handling casualties with impaled, unexploded ordinance (addressed in a 1999 *Military Medicine* paper[3]) needs to be considered.

---

[3]B. Lein, J. B. Holcomb, S. Brill, S. P. Hetz, and T. McCrorey. "Removal of Unexploded Ordnance from Patients: A 50-Year Military Experience and Current Recommendations," *Military Medicine*, Vol. 164, No. 3 (1999) pp. 163–165.

Medical Simulation

- Just starting
- Can train
  - triage
  - interventions, equipment
  - environment
- Over and over

There are benefits in using medical simulations that employ very advanced so-called "smart mannequins" to train medics, nurses, and surgeons in a tactically realistic environment. Your personnel should train on these medically realistic mannequins (that respond in a realistic physiological fashion) over and over so that they get everything just right.

Trauma Research

- Not only rats and pigs, but people
- DoD is essentially prohibited from funding trauma research by restrictive consent regulations
- We can't get better if we can't try new ways
- Need new legislation

Trauma research is critical for the military. Unfortunately, most people don't know that right now the military trauma surgeons are prohibited from doing trauma research on people because of a DoD law that says we must have individual consent. Individual consent for research is not obtainable on trauma patients because by definition they are not able to give consent. The civilians do not have the same restrictions because they can apply the rules of community consent. We need to have community consent applied to military hospitals and military-funded trauma studies. There is no way we can get better in taking care of trauma patients if we don't fund studies that we perform ourselves. We need new legislation.

Surge Capability

- For 18 months the 42 bed CSH was more than sufficient
- Overwhelmed on 3–4 October
  - 36 continuous hours was enough
- Plan for baseline or mass casualty?
- Peacekeeping admissions vs. combat injuries
  - HIV, TB, kids, pregnancy
  - GSW, mine, mortar, RPG injuries,

For 18 months the 42-bed combat support hospital was more than sufficient. However, that 42-bed hospital was overwhelmed by 120 casualties on October 3–4, 1993. Do we plan for baseline casualties or mass causalities? Do we plan for peacekeeping admissions or combat injuries? Very important questions. Hospital equipment and personnel staffing is dramatically influenced by the answers.

FDA Issues

- Fibrin Dressing, rFVIIa
- HSD, Blood substitutes
- There is a risk
  - Doctors that recommend
  - Leaders that say OK
  - Soldiers that are given the stuff
- Recognize the field is different and don't hold us to the US civilian standard
  - 95 vs. 99.9%

FDA issues are critical. Fibrin dressings, drugs that control hemorrhage, different fluids, blood substitutes: all of these techniques, drugs, and products are available. None of them are FDA approved at this point. However, they are all being studied. They are all being utilized in other countries. Right now, lack of FDA approval is preventing some of these very valuable and useful devices from being deployed. Again I think it's important we recognize that the field is different and that U.S. civilian standards do not apply. Our leaders need to stand up and say it's okay to use these things that could and probably would make a difference in the field right now.

Doctrine Risks

- Hypotensive resuscitation
  - Will never be proved in double blinded, placebo-controlled studies with 15 hour evacuation times
- Does that mean we don't do it?
  - No
- Doctrine writers must take risks
  - rapid, relevant, unproved but makes sense

Fresh Whole Blood Transfusion

No crossmatch or testing

Based on Dogtags

Critical to Use

Old information WWI – present

DIC Stops

1/3 of Hospital Donated

50/80 Units given

No transfusions reactions

One of the recurring issues that has influenced every medical deployment that U.S. forces have been on since WWI is fresh, whole blood transfusion. Transfusions relying on no cross matching, no testing, and based on dog tags alone have been done as recently as six months ago in Kosovo. We did it in Somalia. It was done during Desert Storm, in Vietnam and Korea, WWI, and WWII. One third of our hospital in Somalia donated blood. Fifty units were given. There were no known transfusion reactions. Whole blood transfusion is a critical way to help stop bleeding in patients who have lost their clotting factors due to massive hemorrhage.[4] Most young surgeons do not even know how to do this—or that the capability exists. It's easy to do. It's never been taught. Blood bankers don't teach it because it's not FDA approved.

---

[4]S. M. Grosso and J. O. Keenan, "Whole Blood Transfusion for Exsanguinating Coagulopathy in a U.S. Field Surgical Hospital in Postwar Kosovo," *Journal of Trauma,* Vol. 49 (2000), pp. 145–148. R.L. Mabry, J. B. Holcomb, A. Baker, J. Uhorchak, C. Cloonan, A. J. Canfield, D. Perkins, and J. Hagmann, "U.S. Army Rangers in Somalia: An Analysis of Combat Casualties on an Urban Battlefield," *Journal of Trauma,* Vol. 49 (2000), pp. 515–529.

Lesson

- The Medical R&D and Doctrine communities have not recognized this concept as viable. There needs to be a program to devise card-based testing so we will not have to transfuse based upon dogtags, and a predetermined plan outlining how to do the transfusion.

This is very critical information out in the field. There needs to be a program relying on card testing so that we can get better medical information than that provided on dogtags.

Triage Plan

- Starts with buddy care
- Medics
  - Treatment and evacuation order
- Casualty collection point
- Hospital
  - Sick or not sick
- Who does it?
- Who is most important?

Having a triage plan is critical. Triage starts with buddy care and proceeds to medics and the casualty collection point in the hospital. It's very important to teach combat casualty triage and to include concepts that may differ from those in most textbooks.

Lesson

- Triage is difficult
- Sort by the principle of:
  - sick, not sick, and dead
- Medics are the most important component

Medics are the most important component to any triage plan because they are the farthest forward and see the casualties first. They decide who comes back to the surgeons.

Lesson

Expectant

To categorize a soldier to this category requires a resolve that comes only with prior experience in futile surgery that ties up operating rooms and personnel while other more salvageable casualties wait, deteriorate, or die.

Making a U.S. casualty expectant, where he is alive and you're not going to work on him, is one of the most difficult decisions that any surgeon, any medical person, will ever have to make. However, it often times is the correct thing to do. Free up the operating room to take care of others less severely injured but more certain to live.

Hemorrhage Control

Uncontrolled Hemorrhage
- 50% of deaths on Battlefield
- 25% of deaths in the OR

Methods of Hemorrhage Control
- Tourniquets
- Silk Ligatures
- Pressure
- Gauze Sponges
- Clamps

Hemorrhage control is vitally important to saving lives on the battle-field.

Hemorrhage Control

- We are surrounding our soldiers with state-of-the-art technology…
- ...yet when treating hemorrhage on the battlefield we use the same methods deployed for centuries.

Unfortunately, we are currently surrounding our soldiers with incredible technology, and yet the way we do hemorrhage control in the year 2000 is exactly the same way that it was done in the Trojan War, WWI, and WWII. We use the same gauze battle dressing that we have for decades. In the operating room we use gauze dressings, silk ties, and ligatures.

These methods haven't changed for over 2,000 years. This image presents an example of gauze dressing being placed on a wounded extremity during the Trojan War. The following photos document gauze dressings carried by soldiers in various wars. These dressings are unchanged since at least WWI.

INDIVIDUAL DRESSING PACKET
U. S. ARMY
SEABURY & JOHNSON, NEW YORK, U. S. A.
LABORATORIES, EAST ORANGE, N. J.
Contract May, 1917

World War I.

SMALL
BATTLE DRESSING
STERILIZED—CAMOUFLAGED
U. S. ARMY CARLISLE MODEL
(TO OPEN PULL TAB)
RED COLOR INDICATES BACK OF DRESSING
PUT OTHER SIDE NEXT TO WOUND
STOCK NO. 2-396 — LOT NO. 920
DATED: JUNE 30, 1943
HANDY PAD SUPPLY CO.
WORCESTER, MASS.
CONRAY PRODUCTS CO.
NEW YORK, N. Y.

World War II.

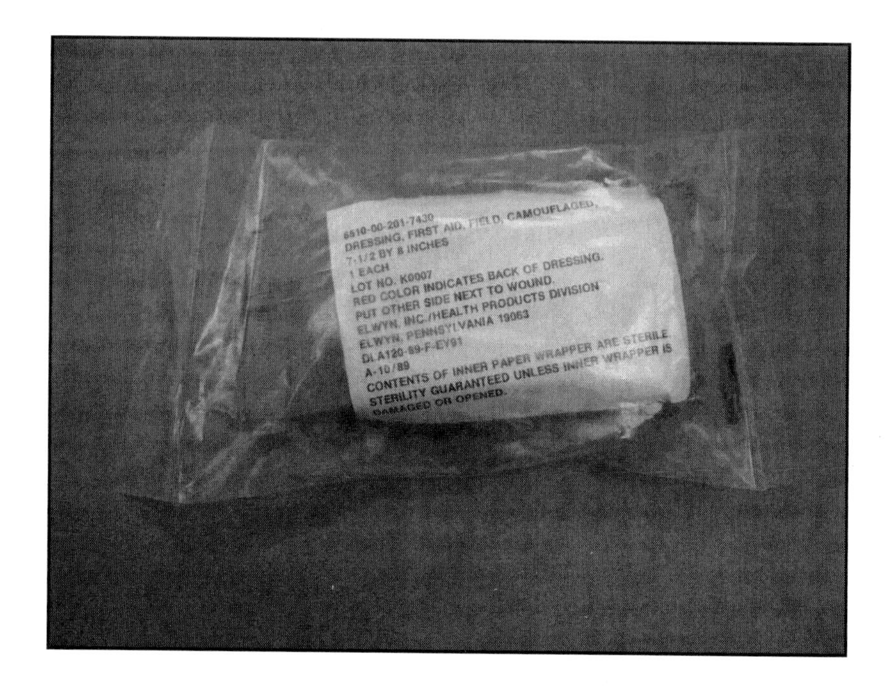

Desert Storm.

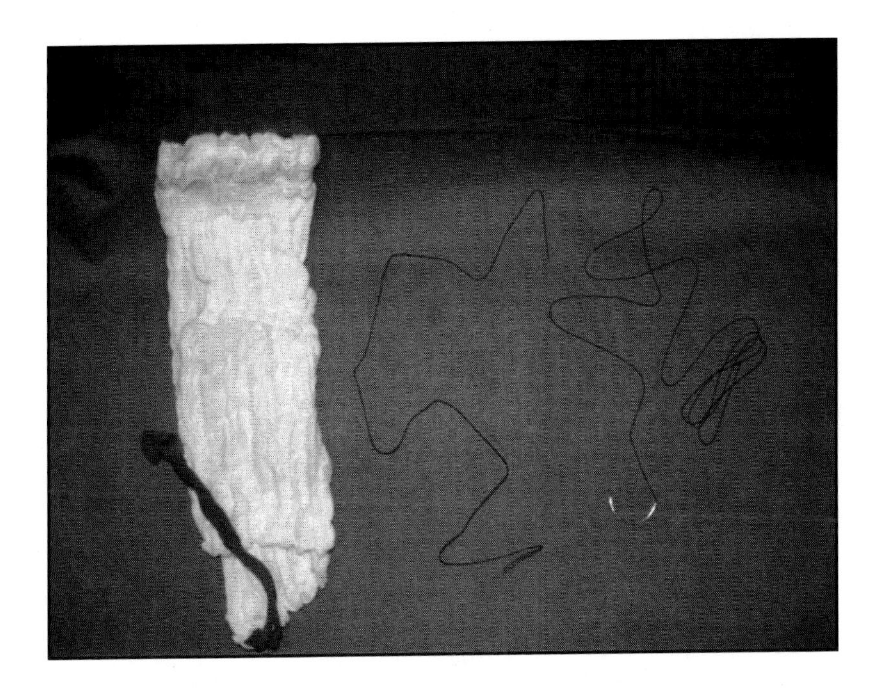

Gauze dressings and silk ligatures currently utilized in the hospital, again essentially unchanged for 2000 years as the primary means of hemorrhage control.

Lesson

- Must improve hemorrhage control techniques
- MRMC is actively funding this research

When you consider that hemorrhage is the number one leading preventable cause of death on the battlefield and the way we do hemorrhage control is 2,000 years old, you understand why MRMC is actively funding hemorrhage control research.

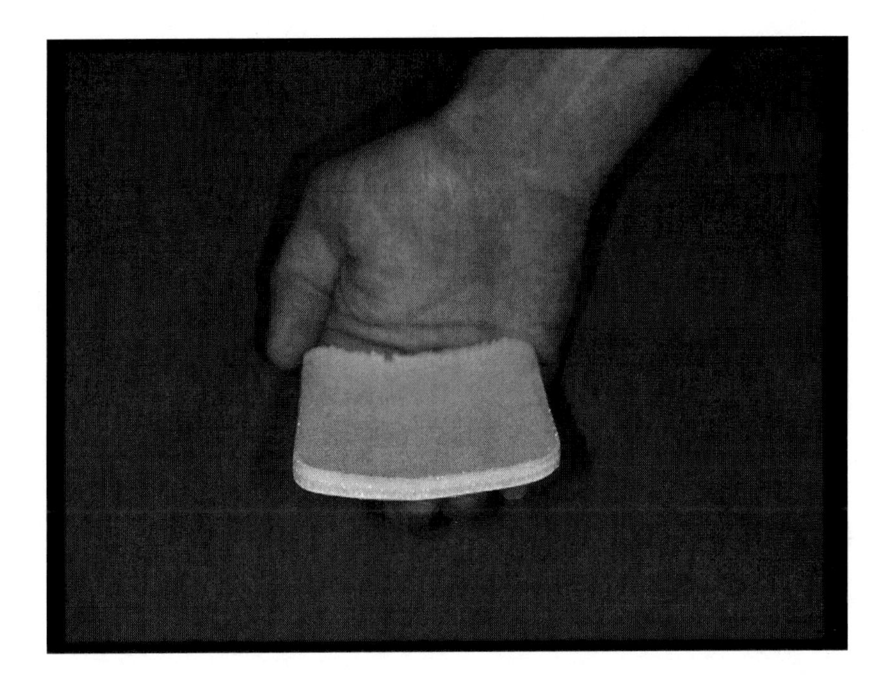

Hopefully in the next year or two we will have deployed a better way to stop bleeding: this Fibrin dressing.[5]

[5] J. Holcomb, M. MacPhee, S. Hetz, et al., "Efficacy of a Dry Fibrin Sealant Dressing for Hemorrhage Control After a Ballistic Injury," *Arch Surg*. Vol. 133, No. 1 (1998), pp. 32–35.

Tourniquets

- Standard issue doesn't work; most medics know this.
- Need a one-handed device
  - –auto inflate?
- Address the 10% of extremity wounds that die from hemorrhage.

Tourniquets: the standard issue army tourniquet that is in every medic's aid bag does not work. Most medics know this. U.S. SOCOM is funding a one-handed device that actually does work. This is very important because 10 percent of those suffering extremity wounds die from uncontrolled hemorrhage.[6]

---

[6]M. D. Calkins and C. Snow, "Evaluation of Possible Battlefield Tourniquet Systems for the Far-Forward Setting," *Military Medicine,* Vol. 165 (2000), p. 379.

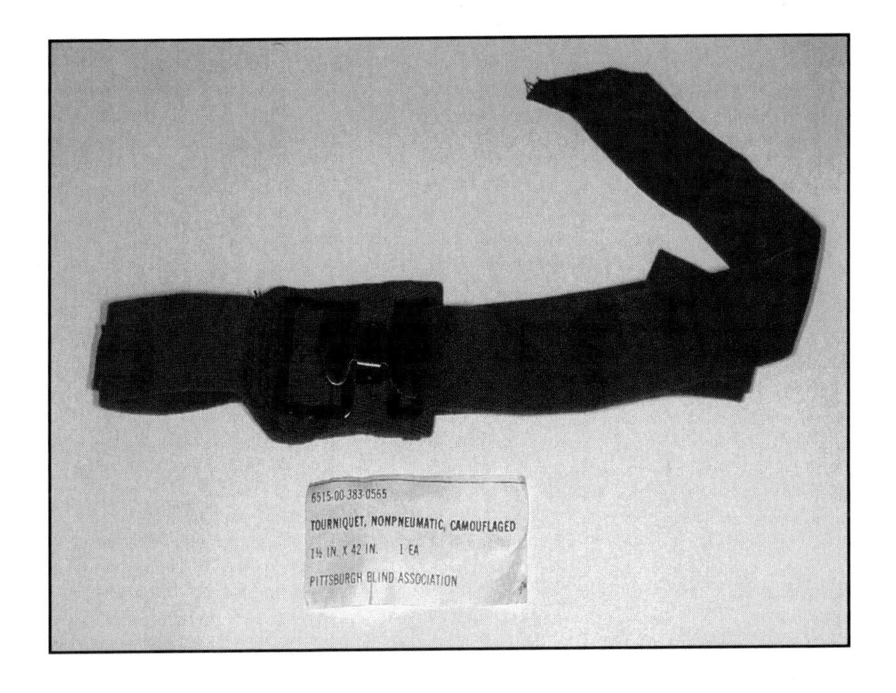

The standard Army tourniquet that most pre-hospital medical personnel feel is not effective.

Lesson

- Need a real tourniquet that works.

The lesson here is that we need a real tourniquet that works and we need it very quickly.

Hypothermia

- 35 degrees C for trauma patients
- Often unrecognized problem
- Significantly increases mortality
- Occurs in the desert

Hypothermia: a trauma patient's getting cold is very important. It's an often unrecognized problem. It increases mortality because the clotting cascade doesn't work when you get cold. It is a problem equally in desert and colder environments.[7]

---

[7]L. M. Gentilello, G. J. Jurkovich, M. S. Stark, S. A. Hassantash, and G. E. O'Keefe, "Is Hypothermia in the Victim of Major Trauma Protective or Harmful? A Randomized Prospective Study," *Annals of Surgery*, Vol. 226, No. 4 (1997) pp. 439–449.

Lesson

- Hypothermia is an independent risk factor for death in the trauma patient. There is not a coordinated plan to prevent and treat hypothermia in combat casualties through the casualty evacuation system.
- Identified in a 1918 JAMA article by W.B. Cannon as a major problem during trench warfare.*

*W.B. Cannon et al., "The preventive treatment of wound shock," *JAMA,* Vol. 70 (1918), p. 618.

Hypothermia is an independent risk factor for death. The military does not have a coordinated plan to prevent and treat hypothermia in combat casualties through the evacuation system. We are essentially no better than we were when Walter B. Cannon wrote in 1918.[8]

---

[8]W. B. Cannon and J. Fraser, "The Preventive Treatment of Wound Shock," *Journal of the American Medical Association,* Vol. 70 (1918) p. 618.

18 *NATURE AND TREATME*

THE PREVENTIVE TREATMENT OF WOUND SHOCK

W. B. CANNON, M.D. (BOSTON)
Captain, M. R. C., U. S. Army

JOHN FRASER
Captain, M. C., R. A. M. C.

E. M. COWELL
Captain, R. A. M. C., S. R.

FRANCE

INTRODUCTION

This paper from 1918 by Walter B. Cannon is from WWI and deals with the preventative treatment of wound shock. It's a classic shock article, widely referenced. What is not talked about is that 75 percent of the article discusses how to keep casualties warm in the trenches.

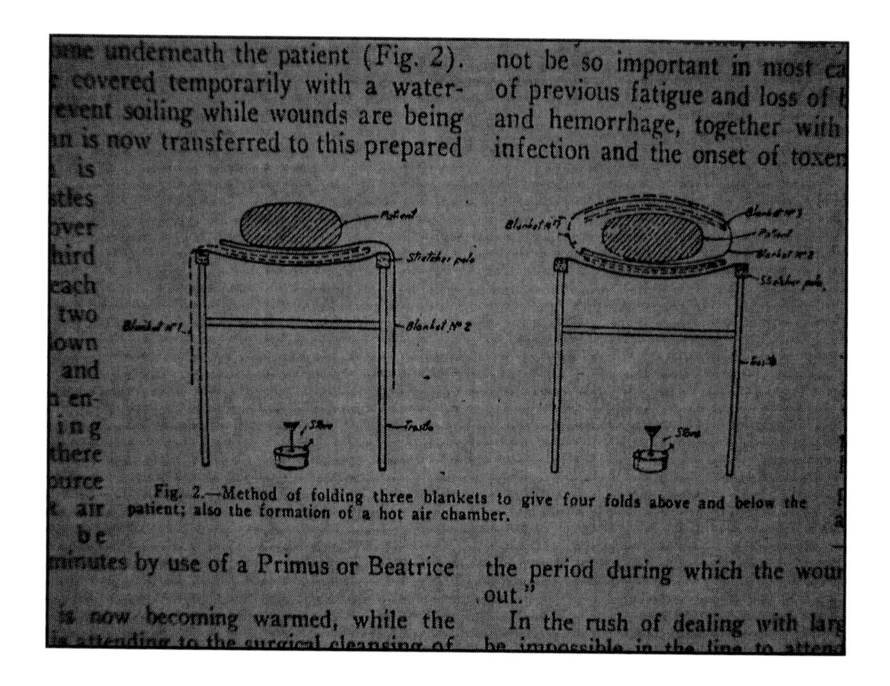

It talks about putting a candle underneath the litter and wrapping the patient in an army wool blanket. We are essentially no more sophisticated today. Most FSTs [forward surgical teams] have no rewarming capability.

Return to Duty Policy

- Unusual everyday decision
- Prevent harm to the soldier and his unit by premature return to duty
- Not a lot of written guidance

Determining a soldier's return to duty status during combat turned out to be a far more difficult challenge than I expected. In the midst of my combat experiences I personally could have done a better job with this. I probably excused three or four soldiers who could have come back to duty and gone back to the firefight. There's not a lot of written guidance on this issue.

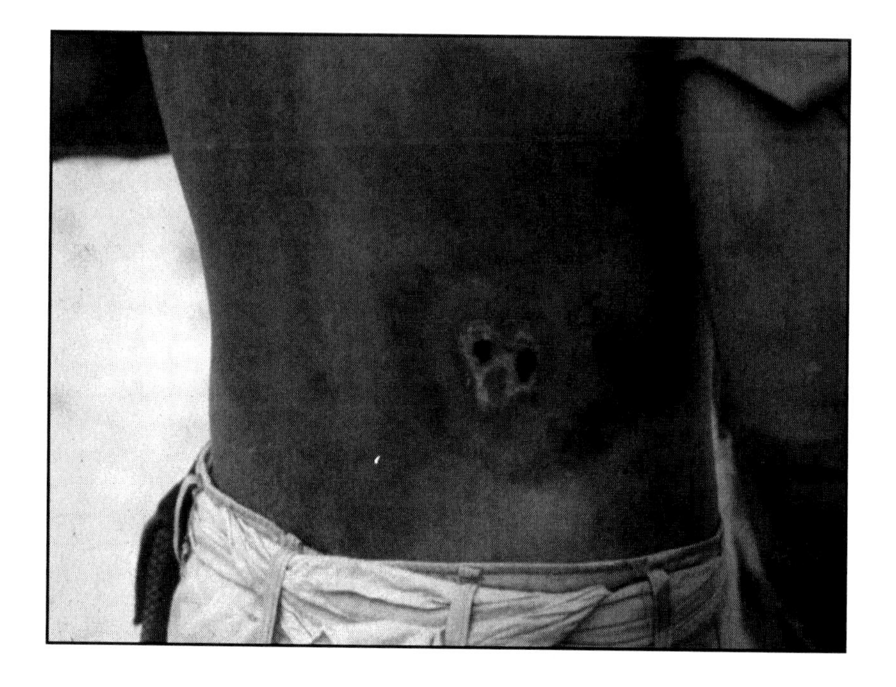

This picture is of a soldier four days after he was hit in the flank by a ricochet round. He came to me with a big bruise. I had him bend over and touch his toes, turn side to side. We checked his urine and sent him back out.

Lesson

- Return to duty as many personnel as possible
- Some could have returned; we were relatively unaware of this concept
- Perception of immediate evacuation for all injuries

However, I did not really have much guidance or experience in this area and probably need to develop more.

MedEvac Problems

Critical Care in the Air

Supply of Physicians, Nurse, RTs

Ventilators / Equipment

Frequency / Availability

Medevac problems are significant. In 1993 we had no critical care in the air; we had to supply our own surgeons to take care of casualties and basically had to fly surgeons and RT personnel out of country with our casualties, leaving our staff further depleted in country while not knowing if we would receive more casualties. Equipment issues between the army, navy, and air force are significant and unfortunately are a recurring problem. This needs to be fixed. We hand off our casualties to people who don't know the equipment with which we are handing them off. We don't know the equipment with which we are accepting them. It makes no sense.

Transferring a patient who was intubated and was post-op from a sternotomy and laparotomy [from the 46th Combat Support Hospital on a two-wheel carrier].

Lesson

- CCATT has fixed this problem
- Make sure the Army / Navy and CCATT (USAF) teams have similar equipment and training to facilitate handoff

The Air Force has actually done a nice job with critical care transport teams that fly in with the airplanes and then fly the patients out with their own equipment.

Multinational Forces

- Reality
  - Not all have equivalent medical care
    - Find out quickly
  - Start medical conferences
    - Establish communication
  - Mass casualty transfer exercise
    - Distribute cases when overrun

Multinational forces sound very nice when you go in, but the reality is that you have varying levels of medical care. It's important to find out quickly who can do what. What we did was to start medical conferences to establish communication among the coalition forces. We quickly learned which hospitals we wanted to send our casualties to and which ones we didn't. We utilized the Swedish hospital extensively when our capacity was exceeded in Somalia.

Data Management

- ID numbers
- X-rays, labs, H&P, op note, plan at bedside
- Telemedicine (read telephone)
- Receiving hospital and research
  - Digital burst
  - Optical record

Data management, keeping track of patients in the middle of mass casualty is very important. Remember that we had 120 casualties for a 42-bed hospital. ID numbers are best put on patients' foreheads with indelible ink; all relevant data (whether it be x-rays, labs, or other information) should be kept at the bedside. Telemedicine is *not* currently useful in the acute management of the trauma patient and in general is highly overrated. However, telephones are very useful, were available, and should have been utilized to call the receiving hospital and speak with the senior physician directly. This would have prepared him for his facility's mass casualty event to be experienced after the transport aircraft arrived. Documentation of injuries for teaching purposes is very useful. But in the midst of mass casualties, I would recommend that we keep this technology on the side. Yet receiving hospitals would obviously greatly benefit from a digitally-burst, video record so that they could review cases for research purposes and set up their receiving areas and staff them appropriately.

Lesson

- Every small hospital in the U.S. has a trauma registry and dedicated personnel to keep track of the patients. They do this to make sure that they are doing a good job and for research purposes.
- The DoD does not have this, so we don't know how we did, and thus we don't have the data to drive improvements for next time.

Every small hospital has a trauma registry and dedicated personnel to keep track of patients. They do this to make sure they're doing a good job for research purposes. The DoD does not have this so we don't know how we did and thus we don't have the data to drive improvements for the next time. This is a real problem and something that we really need to improve on.

Non-Battle Injuries

- Not always sprained ankles and mosquito bites

Non-battle injuries are very important. They're not always minor.

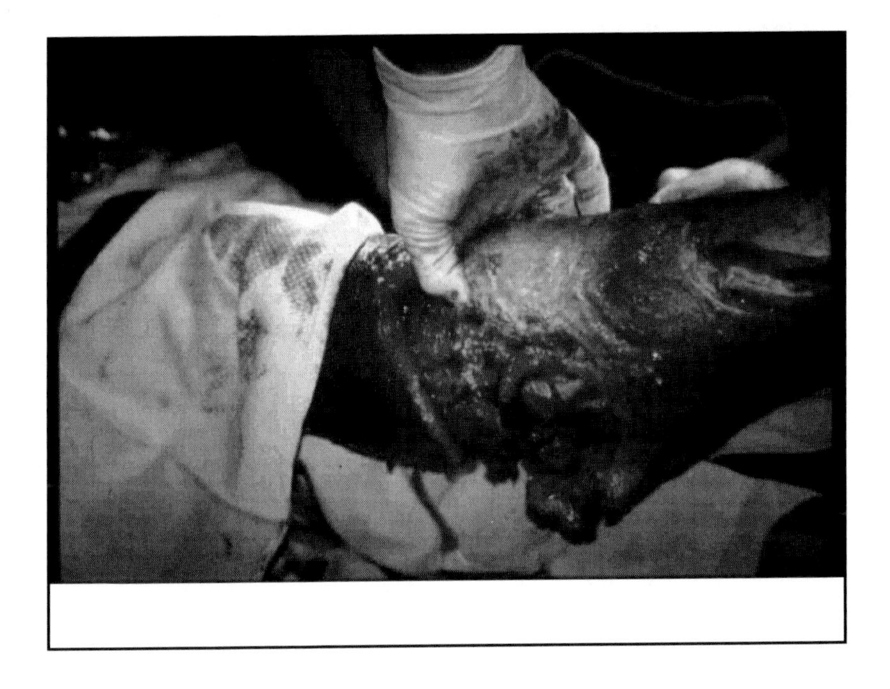

For example, we had a soldier bitten by a shark.

Gadgets and Doctrine
vs.
People

Advanced Technology on the Battlefield

- Diagnosis and Treatment of:
  - Hemorrhage Control
  - Head Injury
  - Airway / Pneumothorax
- Optimal Resuscitation: Level / Drugs / Fluids
- Evacuation
- Rewarming
- Data Capture and Transfer

Gadgets and doctrine versus people: I'm a firm believer in optimal advanced technology on the battlefield. However, we need to direct research to include battlefield concerns. Hemorrhage control, head injury, airway problems, and pneumothorax are preventable causes of death on the battlefield. We need to develop optimal levels of resuscitation with the right fluids, drugs to intervene and help improve hemorrhage control. We need to address the evacuation issues with equipment and timing. We need to address ways to keep casualties warm. We need to address data capture and transfer for research purposes so that we can do better the next time.

Lesson / Challenge

- Realistic Medical Training
- Merge and balance advanced technology with realities of forward deployed troops.
- Demonstrate a need for the advanced technology before committing dollars and time.
- Solve the problems on the current battlefield before moving to the next.

We need to conduct medical training using both field exercises and simulations. We need to focus on trauma, our main problem. We need to merge and balance advanced technology with the realities of forward deployed troops and the problems that those troops have. We need to demonstrate a need for advanced technology before committing dollars and time to developing specific concepts. In other words, let's solve the problems of the current battlefield before moving on to the next.

Training of Medics (Everyone)

- Must act independently
- Maintenance of casualty
- Advanced skills
- Recent trauma experience
  - not just sick call
- Peacetime (Tricare) vs . Readiness
- Equipment and sustainment
- Ask them when you get home

The training of medics: when I say medics I mean everyone from the youngest enlisted medic all the way to the senior surgeon. They must be able to act independently. They must be able to maintain a casualty if they can't evacuate the patient. They must have very advanced skills. They must have recent trauma experience and not just sick call. A lot of our medics have not taken care of a trauma patient recently . . . or ever.

Training

- How do we train for war while in peace?
- Realities of war—Level I vs. Tents
  - numbers, dirt, dark, services, chaos, experience
- Incredible dichotomy

There's a real problem of balancing the realities of peacetime, which means Tri-care medical care, versus readiness training. Doing hernias and breast biopsies will not prepare you for a soldier with his leg blown off by an RPG. Medical personnel need to have the right equipment and to have completed sustainment training with that equipment.

Lesson

- Must have shots, wills UTD prior to deployment
- No requirement for up-to-date trauma skills
- Most DoD MTFs not involved in trauma care…our core responsibility
- We must establish and fund a DoD -wide trauma training standard

In order to deploy we require that every soldier has his shots and wills up to date. There's no requirement for up-to-date trauma skills. Most DoD MTFs are not involved in trauma care, our core responsibility. We must establish and fund a DoD-wide trauma training standard and then enforce that standard and allow our medical people, DoD-wide, to complete trauma training.

Lesson

- Senior leaders must ensure that their hospital personnel are clinically ready.
- Taking care of the locals is an excellent civic action plan and we get our medical system up to speed...so they are ready when our guys get hurt. It also improves morale, exercises the logistic system, and keeps the hospital busy.

One of the ways to do this with deployed forces is allow them to take care of the local nationals. We did this in Somalia through the wisdom of our hospital commander and deputy commander. We took care of Somalis who were injured. That meant that the hospital was ready to roll and knew what to do when our guys came in. That meant our guys were ready to go.

SOF Medics

Treated every casualty correctly

Tough medical decisions

Ran out of supplies

The SOF medics treated every casualty correctly. They made tough medical decisions. They ran out of supplies. They did an outstanding job. They were very, very good.

Nurses

Awesome

Orders on one ICU patient

- keep the fresh whole blood going while he is hypotensive
- draw an ABG and fix the vent
- use any drug in the pharmacy as appropriate
- give some pain meds
- I'll be back in an hour

The nurses at the 46th Combat Support Hospital were awesome. The orders on one of the intensive care unit patients that I gave was, "Keep whole blood going, draw blood gas and fix the ventilator, use any drug in the pharmacy, give some pain medicines." Those were the real orders to one patient and the nurse, a lieutenant, did a very, very good job taking care of that guy.

Two of the nurses of the 46th Combat Support Hospital taking care of a casualty with a laparotomy and a thoracotomy. They did a great job, as did the entire hospital.

Lesson

- This is the level that we need to train for
- The SOF medics are trained for these events and they did well
- If the nurses can accept these orders, then they are ready

This is the level we need to train for. The SOF medics are trained to perform in combat; they did a good job. The nurses can accept those types of orders. They're ready to take care of casualties in the tent.

Definition of a Combat Doc/Nurse/Medic

Young people who must have good hands, a stout heart, and not too much philosophy. They are called upon for decisions rather than discussion, for action rather than knowledge of what the latest writers think should be done.

My definition of a combat nurse, medic, or doctor is out of the Emergency Ward Surgery book: "Young people must have good hands, a stout heart, and not too much philosophy. They are called upon for decisions rather than discussion, for action rather than knowledge of what the latest writer thinks should be done."

Summary

Quality trauma care can be delivered in tents

All providers (doctors, nurses, and medics) must have broad experience

Unique lessons (relearn)

Based upon Military Trauma Database and Medical Center for Lessons Learned

Opportunities exist to improve care in the field

–head injury
–hemostasis
–resuscitation parameters

In summary, quality care can be delivered in tents. All the providers must have broad and recent experience. There are unique lessons that need to be learned and relearned. We need to create a military trauma database and a medical center for lessons learned so that these things don't have to be relearned every time we go in field. Opportunities do exist to improve care in the field, especially in the areas of head injury, hemostasis, and resuscitation parameters.

Mogadishu, Oct 3, 1993

Appendix P

THE URBAN AREA DURING SUPPORT MISSIONS CASE STUDY: MOGADISHU

Applying The Lessons Learned—Take 2

CSM Michael T. Hall, U.S. Army
SFC Michael T. Kennedy, U.S. Army

AGENDA

- Training Focus
- Equipment/Skills
- Collective Training
- Lessons Learned

"…our most probable combat situation—physically grueling, lethal operations encountered in a night, MOUT environment"

Regimental Training Guidance

We know that things that are of real value are in urban areas. The army spends the majority of its time training in a "wooded environment." We believe that this is training to fight the last war, not the next one. Hence, the majority of our training is focused on urban areas. We train as we expect to fight.

Mission

Plan and conduct joint special military operations in support of U.S. policy and objectives.

Who We Are

Light Infantry

How We Fight

Army FIELD MANUAL 7-8
The Infantry Platoon and Squad

This is our mission statement. It is really no different from any other unit in the military. It is very broad, which can present problems as we try to figure out what to focus on with our limited assets, the most limiting asset being time. We will present how we have attacked that problem during the course of this brief.

We are many things to many people, but basically we are light infantry. We do the things that infantry has always done. For that reason we believe this brief is applicable to all units in the U.S. military, not just us. We do have some unique capabilities, but at squad, platoon, and company level there is not much difference between us and other units. We do conduct special military operations, but how we fight is based on standard army doctrine. Field Manual 7-8, *The Infantry Platoon and Squad,* is our bible.

Training Priorities

- **REGIMENT**
  - Train the battlestaff on staff METL
- **BATTALION**
  - Maintain a "trained"("T") status on "Perform Airborne Assault on a Defended Airfield"
- **COMPANY**
  - In the infrared (IR) Spectrum, on Urban Terrain, maintain a "T" Status on Perform Raid
- **PLATOON/SQUAD**
  - In the IR Spectrum, on Urban Terrain, maintain a "T" Status on: Battle Drill (BD) #1, "Platoon Attack"; BD #2, "React to Contact"; BD #6, "Enter Building/Clear Room"; BD #8, "Conduct Breach of a Mined Wired Obstacle"

We have decided that with the myriad of possible missions out there, there is no way we could be good at all of them, especially since we are on an 18-hour string to go into combat. Therefore, we have committed ourselves to being good at a few basic tasks that we feel we could use as the basis for any combat mission. We are convinced that no matter what mission we are given, even on very short notice, we can be successful at it because we feel these basic tasks will cover any combat situation we could face.

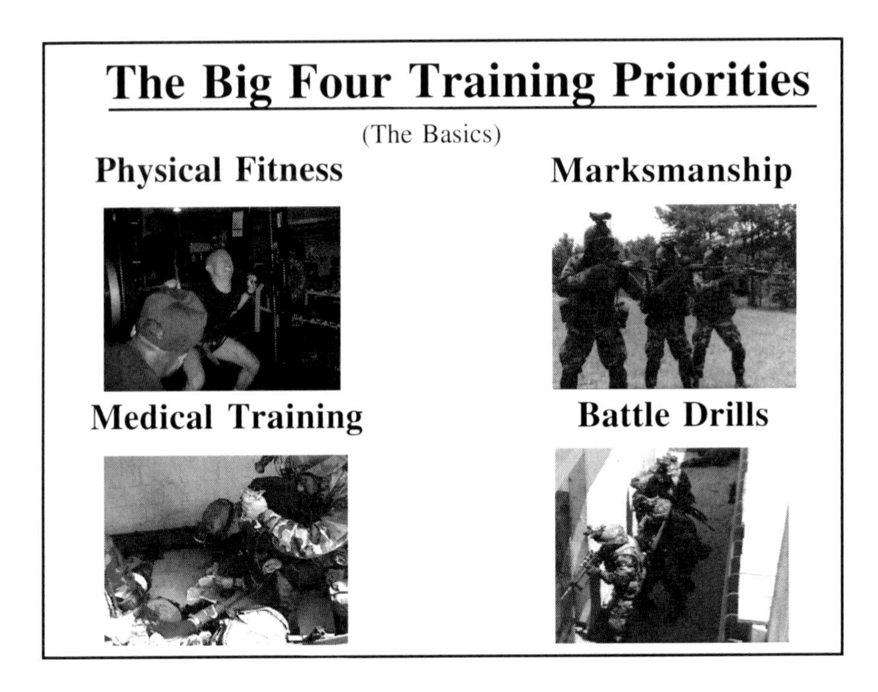

We believe there are four basic areas that a unit needs to master in order to survive and be successful in combat.

Unless we have mastered these four, we won't do anything else. Mastery of these four skills allows us to execute any mission successfully.

Physical Fitness

- Sustain footmarch program
  - 10 Miles Weekly
  - 20 Miles Quarterly

- Combat-focused PT. Formal manual.
  - Bn-level MTTs
  - 90 minutes per day
  - New methods, techniques, equipment

- Combatives
  - Master the 13 core moves
  - Weekly

We know combat, especially in urban areas, will be exhausting. Studies have proven that the better physical shape you are in, the less mental exhaustion you will suffer. Mental exhaustion has proven to be a significant factor in urban combat.

We believe that combat-focused physical training (PT) is one part of training how we will fight. We invested three weeks in a program that involved a squad leader per platoon and a first sergeant per battalion working under the supervision of the army physical fitness school. They developed a program that more closely mirrors the physical activities of combat. This was followed by battalion and company level leader training. The results have been rewarding: fewer injuries, better physical condition, and more success executing combat tasks.

We don't really know if we will have to footmarch long distances in combat; we do know that a sustained foot march program is a tested method to build true endurance in combat soldiers.

The other area in our physical fitness system is our combatives program that we will discuss a little later.

Medical Training

- Individual
  - Every Ranger is a combat life saver
  - Battalion sustainment t raining
  - Regimental-wide standard
- Ranger EMTs
  - One EMT-basic per squad
  - Semi-annual course

- Ranger Medics
  - EMT-P/SOMTC
  - Live tissue training
  - ATLS
- Ranger units
  - Casualty play at all training
  - Mass casualty exercises
  - Realism (moulage, etc…)

Two facts about combat, especially urban: we know people will be injured and wounded and that there are not enough medics. Our ratio is one medic to every 48 soldiers. In order to address this we have implemented a four tier medical system. The first tier: every ranger is a trained combat lifesaver in accordance with (IAW) the army standard plus several additional tasks. We maintain this capability with battalion and company level programs run bimonthly. We are able to maintain a 90+ percent currency rate. The second tier: every squad/section has an EMT-basic certified and current soldier. We maintain this by running semiannual courses. The third tier is the 91B medic at platoon and company level. These medics maintain an EMT-paramedic level of training. This is maintained through internal battalion programs and training at the special operations medical training course. The fourth tier is advanced trauma life support with our battalion surgeons and physician's assistants. Casualty play is incorporated into every training event in order to exercise the capabilities in a realistic environment. It's a big investment, but one that will certainly pay off in combat!

"There are two kinds of people on the battlefield: Marksmen and Targets"

- Marksmanship vs. Qualification
  - PMI
  - Step by Step Process
- The Four Part Program
  - Qualify Day
  - Qualify Night
  - Close Quarters Marksmanship (CQM)
  - Combat (Stress) Fire

That quote sums up combat and the importance of marksmanship. We realized that the standard army qualification standards did not adequately prepare our soldiers for combat. We also realized that the qualification system for developing marksmanship was misleading and did not require soldiers to get better. We broke our marksmanship training and qualification down to the very basics and started over again. We discovered, rediscovered is more like it, that we could get a lot better if we put our time and effort into preliminary marksmanship training without ammo. The information is in the FMs and it really works. The biggest key to success was implementation of weekly dry fire training. Another key was to follow the FMs and train with a step by step method, not moving on to the next step until the standard for the previous step was met: zero, grouping, known distance slow fire, known distance timed fire, practice qualification, and then qualification.

This system seems very time consuming, but it is actually a much quicker way to success. Our marksmanship program has four parts. A soldier must qualify during the day IAW army standards. He must also qualify at night using the day standards. He must then qualify on our close quarters marksmanship table and combat fire lanes, both of which will be discussed later.

Battle Drills

- Focus on the Basics
  - Dry fire
    - Individual movement techniques
    - Team and squad fire and movement
    - TAPE drills
    - Magazine/belt changes
    - Crew drill
    - Rates and distribution of fire control
    - Immediate and remedial action drills
- Attack, React to Contact, Mined/Wired Obstacle, Enter and Clear a Building

We are convinced that if our squads and platoons can fight and win we can accomplish any mission. For this reason, our focus is on small unit drills, battle drills.

It is easy to maintain readiness levels for battle drills. It's the individual tasks that go into them that need constant attention. We have attacked this by executing these tasks in a dry fire mode on a weekly basis, concentrating on those shown here. By maintaining these we get much more out of our collective training and use the limited assets of ranges, ammo, and training time much more efficiently. Retraining time is rarely available because of resources; the typical response is that an event is usually after action reviewed (AAR'd) and the unit walks away with "stuff to work on." That really turns into making the same mistake the next time it is executed. Our approach has all but eliminated that. We are walking away trained by the time we do the culminating event of live firing at night.

The Army FM 7-8 actually has eight battle drills. We know that we cannot maintain a trained status on all eight. By the time you get around to #8, the unit is no longer proficient in tasks 1–4. We cannot afford the building block method due to our being on an 18-hour string. We feel that if we can execute these four, everything is a simple reaction to a situation that is easily executed. The one thing in common with all battle drills is squad fire and movement. If you can't do that, you can't do anything. If your squads can do that, you can do anything. These are all executed in an urban environment.

Ranger Equipment

- Every Ranger rifleman trains and fights with:
  - Night vision goggles (AN-PVS 7/1 4)
  - Squad communication system
  - Ranger body armor, gloves, ballistic goggles, knee/elbow/shin pads
  - RACK (load carry ing system). Wate r, ammo, medical, breaching
  - Medical plus up
  - M4 carbine/M203/M249/M240 with rail system w/M68 and AN-PEQ-2 laser w/IR illuminator, gun light

This is our standard fighting uniform that we use in all environments, leaning towards fighting in an urban environment. In combat, you will not have a chance to go back to the barracks and change this out for that because the situation changed. For that reason we have developed this system.

Individual Close Quarter Skills

- Close Quarter Marksmanship–
  RTC 350-1-2

- Combatives
  - Brazilian Jiu Jitsu

- Urban movement techniques
  - Combat (Stress) fire

Bombs, artillery, long range fires, and the like do a lot of killing on the battlefield. The individual rifleman, squad, or platoon has little control over these. Somewhere between about zero and fifty meters, a soldier does have control regarding whether he lives or dies. We believe this program gives our soldiers the best chance to live. The close fight is a gunfight, just like the Old West.

There will never be a time in the near future where the urban fight won't have the possibility of being up close and personal. Army FM 21-150, *Combatives*, is a wonderful document. The problem is one of focus and time. Maintaining the skills that are described in the FM is impossible. The Gracie techniques (see page 14 of FM) are simple to learn, maintain, and the chances of injury are minimal.

Standard army rifle ranges do a poor job of preparing soldiers to kill the enemy. They do not train him to shoot in the many situations he will face in combat, especially urban combat. We have developed training methods that better prepare soldiers for the situations in which they will find themselves. We call it combat fire, AKA "stress shoots."

CQM—350-1-2 Standards

- Conducted quarterly (25m range)
  - Reflexive firing training
  - Target discrimination
  - Day/night qualification
  - w/wo protective mask
  - Shotgun application
  - Automatic fire

In the close fight, and the urban fight is mostly a close fight for the infantry, who shoots the fastest and most accurately lives. The other dies. We have developed a training program to address this. We call it close quarters marksmanship. The standards are contained in our MOUT training circular and apply to all individual weapons.

Combatives

- Gracie techniques
- Done weekly
- Only 13 core moves
- Builds confidence and aggressiveness
- Qualified instructors
- Competitions
- Pride
- Involves everyone

As mentioned earlier, the technique of combatives we use is a form of jiu-jitsu named after the world champion Gracie brothers. Combatives training is conducted once a week IAW our training circular to maintain skills. We only maintain 13 core moves because we have determined that is all we can retain. What our combatives program gives our ranger more than anything else is confidence and aggressiveness. When the smallest or newest man in the company who has never played contact sports or been in a fight in his life can best the biggest man, he feels pretty good about himself and feels confident that he can beat any enemy. We don't teach knife fighting, rifle drills, or caving in a man's skull with an e-tool. There are just too many things to teach and not enough time to teach them. But we believe the confidence and will to win this program builds are the most important factors. Each battalion, company, and platoon maintains master trainers who are directly trained by the Gracies themselves or by personnel trained as instructors by the Gracies. This validates the program.

We hold regular regimental-through-squad competitions to ensure all understand the importance of the program and to build esprit de corps. The program builds self-esteem, and this attitude rolls over into other combat tasks. Combat is man on man. It doesn't matter who you are. This program has no rank.

Urban Movement Techniques

Combat (Stress) Fire

- Validity, Reliability, Simplicity
  - Assess, refine, improve our combat marksmanship (train as we fight)
  - Closely resemble combat conditions
  - Refine equipment we will fight with
  - Incorporates:
    - Stresses (mental and physical)
    - All static firing positions (standing, kneeling, prone)
    - Moving and shooting
    - Off-hand
    - Shooting over, under, and around obstacles

The culmination of our marksmanship training program is the combat (stress) fire event. This puts together all of a soldier's marksmanship skills and runs him through a course similar to what he would face in urban combat. The fundamentals underlying any such event are validity, reliability, and simplicity. It is designed to assess, refine, and improve combat marksmanship. The event closely resembles the combat conditions to be expected on the battlefield. It is always done with the soldier's full combat load. These events have taught us how our equipment must be modified and how our marksmanship techniques must be altered for maximum effectiveness and "comfort." There is no "standard course." Leaders will set up events based on resources available and the level of training of their men. As long as it incorporates the elements shown here, it is considered a standard.

Progressive Breaching

- Mechanical
  - tools/TTP
- Ballistic
  - Shotgun
- Explosive
  - close proximity breaching

There is no doubt that there will be obstacles on the urban battlefield, and none will be a "standard American-made" door. These obstacles can kill timing, momentum, and soldiers if they cannot be reduced quickly and efficiently. The method we use is called progressive breaching. Before getting to the point that must be breached, the leader must make an assessment of what he is going to do and then prepare. Bullying your way there and then getting "stuck" is exactly what the enemy wants. Assessment and preparedness are the keys to success. Being fully prepared to "ratchet it up" must also be part of the plan. We have three main categories of breaching. The first is mechanical, everything from kicking in the door to using special tools. Each squad carries one set of tools as part of its standard equipment. The prying tool is a 24" Haligan tool common to fire departments. The striking tool is the eight pound short handle sledge. We have developed detailed TTPs for all mechanical breaching methods. This is something that must be trained. It isn't like they show in the movies.

The second category is ballistic breaching using a Remington 870 short shotgun with a special breaching round to defeat locking mechanisms. Each squad carries one shotgun. The 3rd category is explosive, using close proximity charges. We use one standard charge that fits in a cargo pocket. It will defeat any locking or combination of locking systems up to medium strength metal doors. There is no fragmentation danger and the blast overpressure is small enough to allow you to be within a few feet of the door when it blows.

Distracter Devices

- Critical to saving lives
- Non-Lethal (Friend or Foe)
  - Offensive Grenades
  - Flashbangs
  - Simulators
  - Smoke (Signaling and CS)
- Lethal
  - Fragmentation Grenades
  - 40mm
  - AT Weapons
  - Explosive

Before entering a room or building with known or suspected hostile personnel, it is critical to use some kind of distracter device. If not, the enemy will always have the advantage. You can equate the importance of a distracter device to the importance of suppressive fire. It is critical and must be standard procedure. Without it, soldiers will die unnecessarily. There are basically two categories: lethal and nonlethal. Nonlethal are the best choice because they have the same desired effect and there is no danger to friendlies.

High Explosives

(Battalion and Company Assets)

- Critical to success
  - Small arms fire, ineffective/inefficient for killing or suppression
- Must know and understand weapons and munitions target effects
- Critical to footholds. "Breaching" not to blow holes, but to kill. Door is best place to enter.

There is a tremendous amount of good cover capable of protecting a soldier from small arms fire available in an urban environment. So much so that very little killing will be done with small arms fire exchanged between well-trained and matched foes. The use of munitions more powerful than ball ammunition will be critical to success in an urban fight. High explosive ammo is the answer, but not in the conventional sense. Long-range heavy artillery and airborne fire support are not as effective as in a wooded environment. They cannot provide the close-in support needed by the squad, platoon, or company. High explosive ammo is a scarce resource at small unit level because of its weight and bulk, so it is critical that every shot counts. To attack this problem, we conduct events to train our leaders on the true effects of the weapons and munitions at their disposal. The use of high explosive is critical to establishing footholds and the initial breach. Small units will not have anything available to them for creating a "new" hole for assault.

Simmunitions

- Not "paintball"
- Changed our TTPs
- Makes training real
  - Closest thing to two-way live fire
- Opens up many more training opportunities/sites
  - Can "live fire" anywhere

Simmunitions are 9mm plastic rounds that are fired out of a special upper receiver. They can be used force-on-force when certain protective measures are used. Simmunitions is not a paintball game. You know when you are hit and you tend to act as if it is a live bullet. The use of simmunitions has changed the way we fight in an urban environment. It disproved many accepted techniques and validated new techniques like nothing else available short of actual combat. Using SIM is the closest thing to a two-way live fire. There is no cheating and the threat of pain trains soldiers to do the right thing. The paint is washable. The rounds can be used in just about any facility. That means training can be conducted in a variety of buildings and structures that was not available in the past. We have never been told to not come back to any facility where we have employed simmunitions.

Collective Training

- Company Level and Higher is critical
  - Individuals, Teams, and Squads are OK
  - Platoon, Company, and Battalion coordination is biggest weakness in supporting squads
  - Ratio: 1=Squad, 3=Platoon, 1=Company
  - In the hard fight, footholds/entry gained only through using same procedures as breaching a mined, wired obstacle

The conventional wisdom on urban training is that most of the available training time should be given to the squad because the urban fight is a squad fight. That is true. The problem is that the squad needs a tremendous amount of support to be able to close with and destroy the enemy. We are convinced that our individuals, teams, and squads need comparatively little train-up to get to standard. The difficulty is putting it all together so that platoon leaders, company commanders, and battalion commanders can effectively control and focus their assets to support that one breaching squad. This lesson had to be relearned in WWII, Korea, Vietnam, Panama, and Somalia. The urban fight is very much a squad fight, but it requires more command and control than any other contingency. It must be coordinated such that a whole company, and perhaps a battalion, will be concentrating on nothing more than getting one squad across a street.

We think a good ratio of training time is one block of time to a squad, three at platoon, and one at company. It should be realized the squads are training throughout all three blocks.

In the hard fight, the same TTPs used to breach a mined and wired obstacle will be needed to get that initial foothold/entry point. A platoon (-) in support will not work on a determined enemy. We also regularly conduct leader TEWTs downtown and in garrison areas in order to work this task.

Surgical Fire Support

- Lethal/Nonlethal
- Assets
  - Fixed Wing
  - Rotary Wing
  - Organic Mortars
- Urban ROE
- Munitions Selection
- TGT/Observer ID

The very nature of an urban environment limits conventional fire support, especially during the close fight. Nonlethal methods such as jamming enemy communications in order to disrupt his command and control will help. PSYOP and CA operations will be critical to the fight but may only have limited effectiveness due to demographic characteristics and the sophistication of the enemy. Mortars are of some use because of their high angle of fire, quick responsiveness, and ability to fix the enemy. Fixed wing aircraft such as the AC-130 can be very effective because they can get to targets from the top, but they themselves are vulnerable targets. Structures will usually be an impediment to attack aircraft's ability to provide precision support. The rules of engagement will rarely be unrestricted enough to use our largest munitions, those necessary to totally reduce structures. Even if buildings are destroyed, they are usually turned into strong points and propaganda vehicles for the enemy. Munitions will have to be selected carefully and observed fires will be the normal requirement.

"The phrase that pays"

- Shoot till the enemy goes down
- Never move faster than you can accurately engage the target
- Fight the enemy, not the plan
- It's a three-dimensional fight
- Combat patience, be prepared or be dead
- Stacking is for firewood, the 3-second rule
- It takes two
- We kill with HE

We have some simple rules to live by in any fight. They are especially applicable to the urban fight. Shoot until the enemy goes down. The double tap (firing two shots) is not a guarantee. We train using controlled pairs in as many multiples as needed. Speed is relative; you cannot outrun a bullet. We live by the principle that there is only one thing between you and the enemy: your weapon. The enemy doesn't get our OPORD; he doesn't play fair; we only own half of the battlefield at best. Always be prepared for the unexpected; know and understand the commander's intent. The tendency is to focus forward on your next move or where we know or think the enemy is. The urban fight is 360 degrees and forces must be allocated against this. The common tendency in an urban fight is to move fast, maintain momentum, and get across the next street or into the next building or room. Speed can definitely kill in an urban environment. You must be prepared for your next action. Speed and surprise come from preparedness.

The stack is nothing more than a bunch, which we all know gets people killed and reduces the combat effectiveness of a unit down to one man. The team stack is used when entering a room or building, but it is something that is flowed into and never lasts more than three seconds. Fighting a platoon in an urban environment will severely task the platoon leader and the platoon sergeant. The platoon leader must concentrate on his entire area of responsibility, that 360 degrees. The platoon sergeant helps with the internal coordination between squads. The same applies at company level with the executive officer assuming the platoon sergeant role. Everyone has a specific mission in the fight. Their conventional roles must wait until consolidation and reorganization. High explosives are what we use as the primary killer.

Shortcomings in Technology

- Decentralized use of UAV
  - Detect, Deliver, Assess
  - Organic
  - Short Range
- Counter-mortar capability
- Enclosed space, shoulder-fired AT/Breaching Weapon
- Aerial Observer/ FAC(A)
- AP/AT Mine Awareness
- Mobility = Armored Ground Mobility System
  - Force Protection
  - Necessity for ground MEDEVAC platform

These are some lessons learned from recent fighting around the world that are of great help in the urban fight. UAVs can be a great close support asset in assisting our developing situational awareness. A lightweight mortar counter radar capability that is capable of working in an urban environment is needed for the light infantry. Aerial observation greatly enhances the commander's ability to see deep into his battlefield. Booby traps and mines are employed more than ever in an urban environment as an enemy force multiplier. Mobility, especially armored mobility, is important for re-supply and medical evacuation.

Conclusion—75th Ranger Regiment MOUT Brief

- Individual training
  - Standardization of firing tables (25m)
  - Introduction of stress shooting
  - Equipment standardization
- Collective training
  - √ Simmunitions improves force on force to sqd/plt level
  - √ Increased training frequency

Conclusion—75th Ranger Regiment MOUT Brief

- Collective training (cont)
  - Company and higher (STX)
  - Integration of heavy weapons (MG, AT, mortars, attack helos, AC-130)

Bottom Line

The principles of fire and movement apply in the urban environment just as they do in the woods. The cover and concealment is different, the enemy the same, the bullets just as deadly.

Appendix Q

CONCLUDING REMARKS
Dr. James N. Miller
Deputy Assistant Secretary of Defense for Requirements, Plans, and Counterproliferation Policy

Urban Operations: The Road Ahead

RAND MOUT Conference
23 March 2000

Dr. Jim Miller

Deputy Assistant Secretary of Defense
Requirements, Plans & Counterproliferation Policy

If history is a guide, and I think it is, and I think most people in this room would agree, in the coming years and decades we are going to see more operations in urban terrain rather than fewer. And we are going to see American casualties taken. The work of the people in this room, through their pulling together joint, interagency, and international expertise, can make a big difference, a big difference not just in reducing casualties but also in mission success.

Historical Examples

- Peloponnesian Wars (Syracuse & Athens)
  - Punic Wars (Rome & Carthage)
    - Napoleonic Wars (Vienna & Moscow)
      - American Civil War (Richmond, Atlanta, DC)
        - World War II (Berlin, Manila)

- Cold War (Seoul, Hue City, Kabul, Beirut)
  - Post Cold War (Kuwait City, Mogadishu, Sarajevo, Mitrovica, Belgrade, Grozny…)

3/14/01 2

The historical record is clear. Built-up areas have been central to conflict since the time cities were created. The control of cities has been central to success in conflict. Wick Murray, an historian, has written a nice short paper summarizing the role of cities in 19th-century and 20th-century conflict. Because most of you concur regarding the importance of cities I won't go into the details of these conflicts.

In preparing to speak, I reviewed something I hadn't looked at for 20 years: Thucydides' *Peloponnesian Wars.* One might be tempted to draw an analogy between the over-stretched great power Athens and the United States today. However, the analogy is very imperfect. One difference offers insight into urban operations for the United States. Athens was attempting to impose its will by coercion— ironic as that is considering that it was the birthplace of democracy. What the United States is trying to do in its national security strategy is support human rights, support democracy, and support free trade. The fact that we hold these values, and that we are attempting to uphold these values internationally, imposes serious restrictions on how we conduct conflict—particularly on urban terrain.

Office of the Secretary of Defense

Why Cities Are Important

- Political, Economic, Psychological Centers of Gravity
- Key Logistical and Operational Hubs/Landscape
- Possible Sanctuary for Adversary Forces
- "Because That's Where They Keep the People"

U.S. Military Dominance in the Open but

Significant Limitations in Cities

4/27/01 3

Cities have been important in the past. They are important today, and they will be at least as important in the future. With U.S. dominance in open terrain, more opponents are going to look at moving conflicts into the cities.

Cities hold most of a nation's worth and logistical support. They are a center of gravity. However, that statement doesn't capture the full importance of urban areas. Cities are a central focal point for national and ethnic identity. Think about what it means if you're fighting for your city, for your country, for your tribe.

You may recall the famous bank robber Willie Sutton. When asked why he robbed banks, he said, "Because that's where the money is." In part, that's why conflicts will occur in the cities, because that's where the people are. A significant number of future operations are going to be humanitarian. To deliver that aid you have to go into the city, where the people are. Peacekeeping operations will require going into the cities to protect the people. Disaster relief operations will require entering cities to restore normalcy for the people there.

Even during major theater wars we could expect to find significant conflict taking place in cities.

The "CNN effect," which was coined around the time of Mogadishu, will only increase with globalization, including the growth of the Internet. That trend is going to make it more difficult to win urban conflicts at the strategic level.

Office of the Secretary of Defense

Some Urban Missions

- Destroy Key Targets in a City (Baghdad, Belgrade)
- Peacekeeping (Mitrovica)
- Capture a City (Kuwait City, Grozny)
- Defend a City (Seoul, Srebrenica)
- Humanitarian Assistance (Port-au-Prince)
- Point Defense (Ports/Airfields, Embassies)
- Point Offense (WMD sites, Raids, NEOs, Rescues)
- Civil Support (LA Riots, Olympics, Seattle)

3/14/01 4

Here are some of the key missions we have found ourselves executing in the past. Together they amount to the full range of military conflict. Everything we find ourselves doing, we'll be doing in cities.

At the bottom of the chart you'll notice that not all of these missions have been done outside of the United States. Some lessons learned from urban operations are applicable to operations in which the military support civilian authorities in American cites. The National Guard in particular has taken on a number of roles, to include disaster relief and WMD consequence management.

All of the missions on the chart share a few features:

- Interagency planning and execution are required for success, and all but that last category require coalition operations.
- All of them put a premium on rapid and decisive response.
- All of them put a premium on force protection and minimizing noncombatant casualties to win at the strategic level.

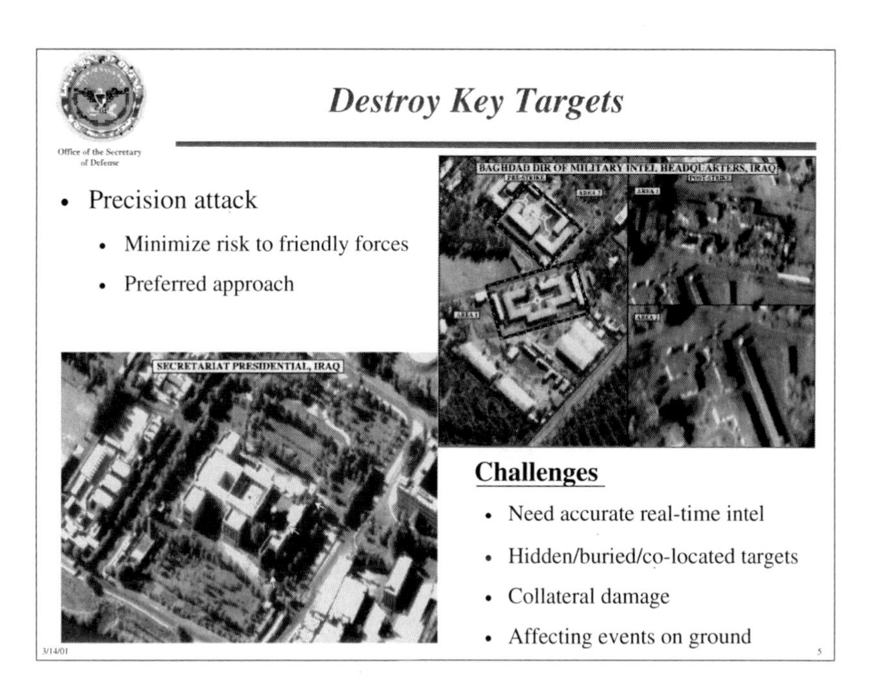

I'm going to go through a few of the missions to show where we stand today and what some of the difficulties are.

Precision Attack is a good way to minimize both friendly and non-combatant casualties—when it is in fact possible. If it can be done, this is obviously the preferred approach. Shown here is a bomb damage assessment image from Operation Desert Fox (Iraq), where we took out some command-and-control facilities and had limited success in taking out a WMD facility.

Aerial attack—and even .50 caliber sniper rifles—have significant challenges and serious limitations in urban terrain. You often need boots on the ground to be successful in applying precision force, and to seize and hold terrain. Rules of engagement and the desire to minimize noncombatant casualties limit the use of lethal force.

There are some new concepts for applying precision force in cities, and this is an area where further work is warranted. I recommend to you the recent RAND paper[1] that explores new concepts for air operations in urban terrain.

---

[1] A. Vick et al., *Aerospace Operations in Urban Environments: Exploring New Concepts,* Santa Monica, CA: RAND, MR-1187-AF, 2000.

Peacekeeping

Challenges

- Combatant/Noncombatant Mix
  - Timely threat ID
  - Minimize innocent casualties
- Risks to friendly forces
- Coordination with NGOs/ PVOs/local authorities
- Maintaining support (local, domestic, int'l)

3/14/01 6

Here we see American soldiers in Kosovo doing door-to-door weapons searches. The women do not seem overawed; that's a good sign when you're doing peacekeeping operations.

There are significant differences in how the different contingents in Kosovo are conducting their peacekeeping operations. How each does force protection and how proactive each is at preventing conflict provides lessons for the others. A continued international effort outside of real world operations, i.e., concept development and experimentation, would facilitate overcoming these challenges.

Office of the Secretary of Defense

Capturing a City

Challenges

- Force protection
- Combatants hiding among innocent
- Defeating adversary without destroying the city—and adding to strategic problem
- Maintaining popular support

3/14/01 7

Capture a City

These pictures are from Grozny, the lessons of which are still being debated. We've seen a big shift in Russian tactics from 1994 to this latest conflict.

There were significant casualties on both sides. Some of you may have seen the recent news reports for the funeral of 86 Russian paratroopers in Moscow. The CNN effect has been suppressed in Grozny, but it has had some effect in Russia.

If you look at the Russian operations in Grozny you'll see that there were significant civilian casualties and much of the city was destroyed. It is seriously open to question as to how long-term Russian national security interests are being advanced and how this will play out.

There is an opportunity to work with the Russians and learn from each other.

The slide shows the result of one failed attempt to defend a city because of a lack of capabilities. At the bottom of the slide is a picture of General Krstic on trial at The Hague.

The first lesson is to avoid getting into a situation like that of the Dutch peacekeepers: outnumbered and outgunned. It would be useful to think about how to defend against a larger force that is technologically inferior. Nonlethal weapons may hold some promise. Lethal weapons would have certainly been called for in this case.

One could also think about a small force defending against a technologically inferior population, as was in part the situation in Mogadishu.

If we and our allies are more prepared in the future, then perhaps we will be less likely to have to look again at images like this.

Office of the Secretary of Defense

Challenges (A Partial List)

POTENTIAL FOR STRATEGIC IMPACT

- Restrictive ROE
  - Risks to U.S./Allied Personnel
  - Intermingling of Noncombatants
- Coordination with NGOs/PVOs/Local Authorities
- Geospatial Representation & Navigation
- Urban Intelligence Collection/Dissemination
- Communication in Urban Canyons
- Non-Lethal Weapons Issues

A Daunting Challenge Nobody Wants

3/14/01 9

This is just a partial list of the challenges in conducting urban operations.

The bottom line is that urban operations are tough to do, whether during peace operations or warfighting. There will be a serious risk of casualties to all parties involved, but it's a job that we will be doing in the future and will have to do well.

A comment on non-lethal weapons, which some see as a panacea. I think they have some real potential if we can work through the tricky policy and legal issues involved. I would like to encourage those involved in that to press on. Through field experimentation, including policy and legal reviews, we can work through some of the more difficult issues.

The focus of this chart is on the tactical and operational levels. However, each of the challenges listed could cause something to happen very quickly at the strategic level.

The CNN effect, globalization, and the rapid dissemination of information and images are going to be more important in the future. We think of Operation Allied Force as an air war. The Washington Post ran a series on the conflict that featured 27 photographs. Five photos were about diplomacy and showed decision makers. Five were on airpower and pictured aircraft. The rest were of people suffering.

The strategist saw the conflict in geopolitical and national interest terms, looking at issues such as the coherence of NATO and the effectiveness of airpower. But a lot of people—most people in the U.S. and likely overseas—saw the conflict as a human interest story rather than a national interest one. That fact has strategic implications and is essential to understand if one wants to win at the strategic level.

National Defense Panel

- Findings
  - Increasing likelihood of operations in cities
  - Difficult conditions include noncombatants, skyscraper "jungles"
  - Possible contingencies include targeting and strike, urban control, urban defense, eviction operations
  - Make every effort to avoid unilateral urban operations
- Recommendations
  - Expand research on urban warfare
  - Establish a Joint Urban Warfare Center

4/27/01 10

This chart shows what some outside of DoD, the congressionally mandated National Defense Panel in 1997, have had to say about our efforts in urban warfare. The NDP was trying to be very forward looking, out to 2025. It had a section on space warfare and a section on missile defense. You can see their recommendations relating to urban operations.

One of the key conclusions was that urban warfare was going to be more of a problem in the future and that the DoD needed to take significant coordinated action to get better at it.

US Commission on National Security
Vol. 1: Future Threats

- Fragmentation or failure of states will occur with destabilizing effects on neighboring states
- Foreign crises will be replete with atrocities and the deliberate terrorizing of civilian populations
- The United States will frequently be called upon to intervene militarily in a time of uncertain alliances

"The emerging security environment in the next quarter century will require different military and other national capabilities."

4/27/01 11

The next major external review of the Department is underway: The Hart-Rudman Commission. It has produced the first part of its report that looks at future security issues and future threats.

Shown are three of the twelve principal recommendations from the first part of the report. Each is directly relevant to urban operations. One of several key conclusions appears at the bottom, which I think also has direct application to urban warfare. It really is going to require the effective integration of all elements of American national power as well as effective interagency and coalition operations.

Office of the Secretary of Defense

General Accounting Office Review
February 2000

- SECDEF should designate lead for MOUT
- Designated lead should:
  - Develop DOD-wide strategy
  - Expedite development of training standards
  - Estimate required resources
  - Establish priorities
  - Examine integration of joint experimentation
  - Develop game plan for facilities
- Need to determine MOUT intelligence requirements

DoD agreed

3/14/01 12

Now, going from the strategic level to the level of the bean counters. The GAO released a report in February; you can see some of their findings here. DoD agreed with the GAO, with a few minor qualifications.

GAO noted a lot of positive efforts going on: significant Army and Marine Corps efforts, the work of the Joint Staff's Joint Urban Working Group and the considerable progress it has made. What GAO found lacking were resources and a focal point for allocating those resources.

Some Emerging Capabilities

- Accurate, Up-to-Date Digital Maps
- Real-Time Fused ISR for the Urban Battlespace
  - e.g., UAVs, robotics, unattended sensors
- Navigation Aids
- Minimal Collateral Damage Weapons
- Improved Non-Lethal Capabilities
- Secure, Reliable Comms
- Technologies for Improved Force Protection

Questions:
- **How quickly will these capabilities emerge?**
- **How well will they be integrated to meet operator needs?**

3/14/01 13

New technologies can help, even if they can't solve the problem on their own, or in concert with new techniques.

The Marine Corps, Army, and SOCOM all have significant efforts under way to try out these new tools and operational concepts. I think it's significant that the Air Force has for the first time recently been involved in experimentation to improve its capabilities in urban operations. The Navy, in one of its most recent fleet battle experiments, supported the Marine Corps during Urban Warrior. So we're moving to better joint concept development and experimentation.

We need to exploit new technologies—and the good work done on the ground by operators. This calls for an integrated joint concept development and experimentation activity to identify the most promising technologies, how can they be exploited for new operational concepts, and to set priorities. A lot of that is going on at the squad and platoon level today. But by integrating various parts of the Department's efforts we can move faster and more effectively.

A second thing we need is a focus on the problem from the perspective of the joint force commander, one that brings together all the component capabilities, agencies, and international partners. As earlier briefers have shown, we saw this work well in some parts of Somalia but work not so well in other parts of that country. The failure to look at what tools the joint force commander needs is a critical missing piece in DoD efforts today. I think we are close to moving ahead on that, as the next slide will show.

Build a Roadmap to the Future

- Establish Baseline Capabilities
- Identify New Concepts and Emerging Capabilities
- Build on Service, Other Efforts
  - Concept Development and Experimentation
  - Investment in ISR, communicati ons
  - Non-lethal and other technologies
  - Doctrine and training
  - Facilities
- Establish Key Mission Needs and Deficiencies
- Plan Interagency, International Outreach

ENDSTATE: A Plan for Significant Improvement in Joint Capabilities

3/14/01 14

The Joint Urban Working Group has laid the foundation for an integrated DoD-wide roadmap. That effort is now underway and being led by the Joint Advanced Warfighting Program. You can see some of the key goals it means to achieve.

The roadmap approach is very simple. Look at what we are doing today and how we are doing it. Establish what the baseline capabilities and concepts are. Identify shortfalls and opportunities, and then prioritize efforts to get better across the board: acquisition, concept development, experimentation, research & development, and then, most importantly, identify new operational concepts.

I want to emphasize the importance of interagency and international participation that includes NGOs and PVOs, or at least proxies for them. Future operations are going to involve them; we need to train with them.

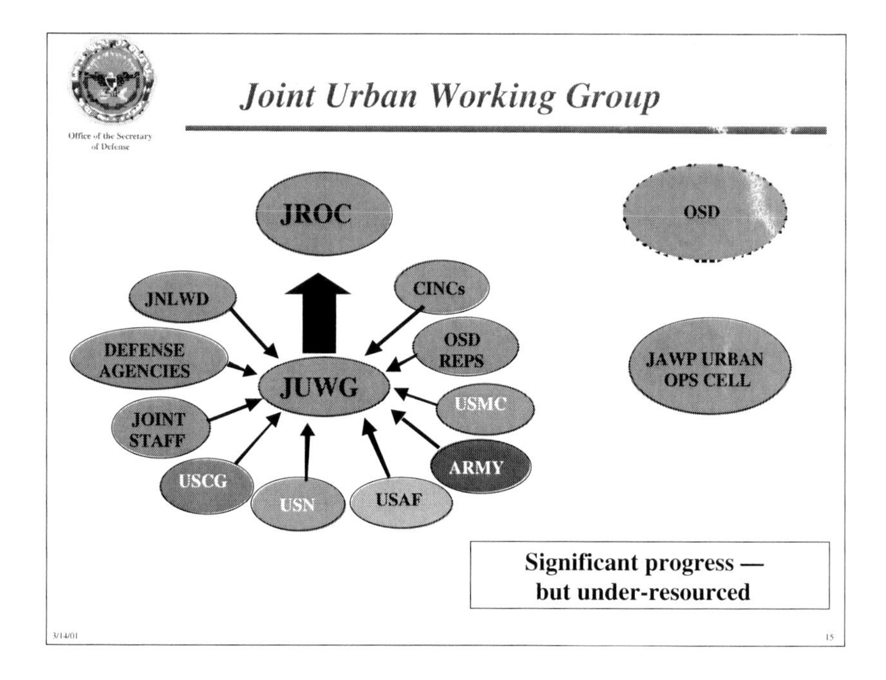

The Joint Urban Working Group reports to the JROC. You can see some of the members arrayed around the JUWG. The JUWG was set up in May 1998.

I think it's incredible what they have accomplished given limited resources. The JUWG has spurred joint doctrine development, and in the meantime developed an operational handbook. It is now reviewing mission needs. It has identified a number of relevant requirements, to include modeling, concept development, and experimentation. That's significant progress.

But it's clear to me that the JUWG has been under-resourced and that we need a focal point in the DoD with greater resources. The JUWG has been appropriately focused at the tactical and operational levels, but I think we need an OSD effort focused on the strategic level, one that drives interagency and international cooperation on both the military and policy sides. Two groups stand out here as shown on the right of the slide: OSD and the JAWP Urban Ops cell, which I'll talk about next.

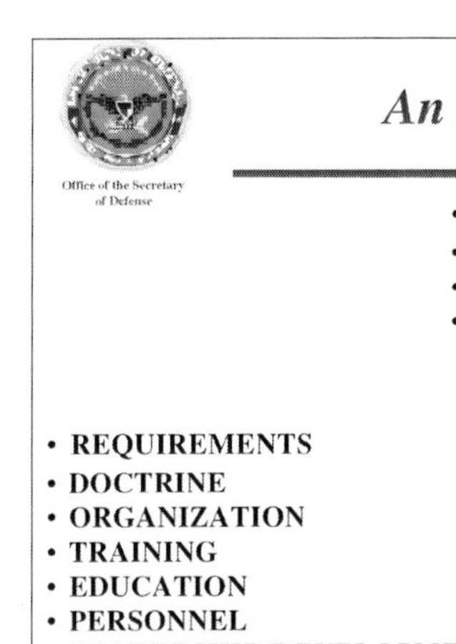

I'll start with the bottom line. We need a single, authoritative focal point in the DoD. I'm not talking about a czar that tells the services what to do, but rather a group with sufficient resources to accelerate joint doctrine development, joint concept development and experimentation—one that can really work the problem from the joint force commander's perspective. His needs are critical in developing new technologies, new concepts, and experiments.

The Joint Urban Working Group has done a great job, but it has had to work with extremely limited resources. The path ahead for the near term is fairly well set: the JAWP is working on a DoD roadmap.

An OSD UWG, as shown on the right in the slide, doesn't really exist today, and I'm not sure if it will exist tomorrow either. Currently we are not optimally organized for this. It's the OSD urban operations champion that we need to cut across all the areas within and outside of OSD.

Office of the Secretary of Defense

Next Steps for DOD…

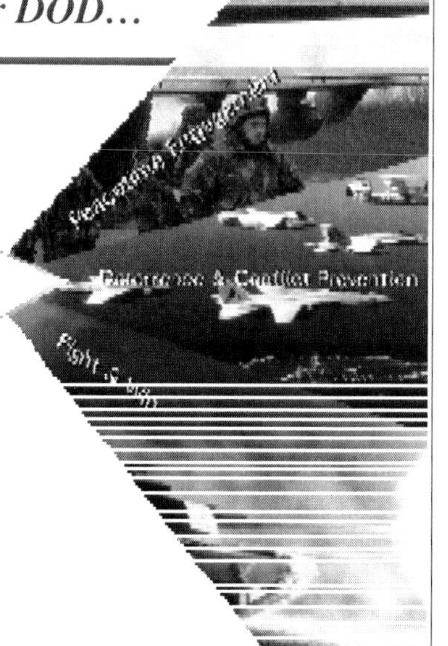

- **Roadmap** to synchronize activities, set priorities
- **Training** standards, tools, and facilities
- **Joint Experimentation**
  - Investigate New Concepts & Capabilities
  - Leverage New Technologies
  - Conduct Rigorous Red Teaming
  - Recommend Changes in DOTMLPF
- **Integrated effort** —joint, interagency, and multi-national—to improve capabilities

3/14/01

The next step for DoD is going to be hard. It's going to be hard to get the resources necessary to set up a focal point, develop new technologies, and explore new training approaches. It's going to be hard to deal with the view of some in DoD that we shouldn't be going into cities and shouldn't waste resources improving our capabilities to do so. That view is not a rare one in the department.

The good news is that the Joint Urban Working Group has gotten the department started. There are some impressive activities going on in the services, particularly in the Army and Marine Corps. The amount of joint work is growing and we are developing a roadmap that will synchronize the department's activities.

We are close to having a critical mass of knowledge and people working on the problem to move things ahead.

The Road Ahead

Think

- **Joint, Interagency, Multinational**
- **Tactical, Operational, Strategic**

Work

- **As part of a Team**

3/14/01 18

The future will bring American forces into conflicts in cities. We will have casualties. Our success in the operation overall, at the strategic level, will depend on how well we do in the cities. That is going to be a big challenge.

The urban problem won't go away. At the tactical level we have some new approaches to improve our capabilities, but at the strategic level the problem is going to get harder and harder for the reasons I've talked about.

No single person, service, agency, or organization has the complete answer. But the prospects for future improvement in capabilities are good over the next few years given the cooperation of all the parties involved.

Thanks to Russ Glenn and to RAND for sponsoring this important conference. And thanks go to all of you in the room who are working to improve our capabilities for urban operations. Keep pressing hard.